# LIFE OF THE PAST

James O. Farlow, Editor

# DRAWING OUT LEVIATHAN

*Dinosaurs and the Science Wars*

KEITH M. PARSONS

*Indiana*
*University*
*Press*
BLOOMINGTON AND INDIANAPOLIS

This book is a publication of
Indiana University Press
601 North Morton Street
Bloomington, Indiana 47404-3797 USA

http://iupress.indiana.edu

Telephone orders   800-842-6796
Fax orders   812-855-7931
Orders by e-mail   iuporder@indiana.edu

© 2001 Keith M. Parsons

All rights reserved

No part of this book may be reproduced or utilized in any form or by any means, electronic or mechanical, including photocopying and recording, or by any information storage and retrieval system, without permission in writing from the publisher. The Association of American University Presses' Resolution on Permissions constitutes the only exception to this prohibition.

The paper used in this publication meets the minimum requirements of American National Standard for Information Sciences—Permanence of Paper for Printed Library Materials, ANSI Z39.48-1984.

Manufactured in the United States of America

**Library of Congress Cataloging-in-Publication Data**

Parsons, Keith M.
Drawing out Leviathan : Dinosaurs and the science wars / Keith M. Parsons.
   p.   cm. — (Life of the past)
Includes bibliographical references and index.
ISBN 0-253-33937-5 (cl : alk. paper)
1. Dinosaurs. 2. Science—Philosophy. 3. Science—Social aspects.   I. Title.
II. Series.

QE861.4 .P37 2001
567.9—dc21

2001016803

1  2  3  4  5    06  05  04  03  02  01

To the memory of the one who taught me to love

knowledge for its own sake:


Arthur W. Parsons, Jr. (1920–1999)

Canst thou draw out leviathan with an hook?
    Or his tongue with a cord which thou lettest down?
Canst thou put an hook into his nose?
    Or bore his jaw through with a thorn?
Will he make many supplications unto thee?
    Will he speak soft words unto thee?
Will he make a covenant with thee?
    Wilt thou take him for a servant forever?
—Job 41: 1–4

# CONTENTS

Acknowledgments
*ix*

Introduction: Why the Science Wars Matter
*xi*

**1**
Mr. Carnegie's Sauropods
*1*

**2**
The Heresies of Dr. Bakker
*22*

**3**
The "Conversion" of David Raup
*48*

**4**
Are Dinosaurs Social Constructs?
*80*

**5**
*Le Dinosaure Postmoderne*
*106*

## 6
### History, Whiggery, and Progress
*126*

## 7
### Beyond the Science Wars
*150*

### Notes
*177*

### References
*195*

### Index
*207*

# ACKNOWLEDGMENTS

At age thirty-eight I returned to graduate school as what was euphemistically referred to as a "nontraditional" student. A previous Ph.D. (in philosophy) had gotten me a succession of part-time and temporary positions . . . and no hopes of tenure-track employment. In the fall of 1990 I entered the University of Pittsburgh's formidable program in the History and Philosophy of Science. In my first term I encountered James G. Lennox in the introductory seminar in the history of science. He gave me a B+, which was quite a shock after finishing with 4.00 averages in my previous two graduate philosophy programs. Naturally, this earned my respect and motivated me to a higher level of performance. When I made an A in Jim's class on Aristotle's biology, this was a proud achievement.

I asked Jim to supervise my dissertation and was gratified when he accepted the role. He was an excellent supervisor, unyielding when he perceived carelessness but quick to encourage a job well done. Since this book grew out of the dissertation, Jim is the main person to be thanked. The other members of the dissertation committee, Professors Ted McGuire, Robert Olby, and Harold Rollins, also earned my sincere appreciation. Finally, Hugh Torrens read the doctoral dissertation which served as the basis for this work and made many valuable recommendations.

Part of this book is based on work done as a research associate at the Carnegie Museum of Natural History in Pittsburgh. Paleontologists Chris Beard and Mary Dawson of the Department of Vertebrate Paleontology were great help and wonderful companions while I did my research at the Carnegie. Chris took me to dig fossils with him in Wyoming—literally "hands-on" experience. Elizabeth Hill, collection manager for the DVP, was most helpful in guiding me through the archives. Thanks also to Lance Lugar, paleontologist and head of the Physics Library at Pitt, for his bibliographical help and our enlightening informal conversations.

I would like to thank my friends and colleagues at the University of Houston, Clear Lake, for their interest and support. Without the security and stability of a "real" job with good friends and congenial

## Acknowledgments

colleagues, this project would have been impossible. Thanks also to two persons whom I shall not name, one a philosopher and the other in cultural studies, who read portions of the manuscript and found them highly antithetical. I am sorry to have caused them pain, and though I have tried to justify or qualify some of the things they found inflammatory, I am sure the final product will still not please them. Their responses showed me just how polarizing the science wars have been and inspired me, in the final chapter, to offer an irenic suggestion.

I would like to thank Bob Sloan and the staff of the Indiana University Press for their encouragement and hard work on this project. A few years ago someone asked me to recommend a top dinosaur paleontologist. Without hesitation I replied that in my opinion James O. Farlow was among the best. It came as a most pleasant surprise to discover that Jim Farlow would be one of the readers of my manuscript. I would like to thank him for his support of my initial proposal and his subsequent work on my behalf. I would also like to thank M. K. Brett-Surman for reading my manuscript and for many helpful comments and suggestions.

Some of those involved in research on the issues discussed in this book may be chagrined to find that their work is barely mentioned, if at all. This book has a multidisciplinary topic, crossing the fields of philosophy, history of science, sociology of science, and paleontology. So wide-ranging a discussion cannot pretend to completeness. Also, this work is not intended as a survey of all the relevant research or opinion nor is it meant to bring the reader up to date on all the latest findings or discussions. Scientific views are mentioned, not so much for their own sake, but as illustrations of the processes of science. Whether these views are the latest word, or have since been qualified by their authors, is not my main concern.

Finally, thanks always go to Carol, for her understanding and support, to my mother, Charlotte Parsons, my sister, Kay Beavers, my brother-in-law, Danny, my niece, Erin, and my good friend, Becky. My father always encouraged my interests, from my boyhood fascination with dinosaurs to the writing of this book, which he did not quite live to see finished. My thanks to him are expressed in this book's dedication.

# INTRODUCTION
# WHY THE SCIENCE WARS MATTER

LIKE SO MANY of my generation, I acquired an early fascination for dinosaurs. Movies, television, and comic books of the 1950s and 1960s treated us Baby Boomers to exciting images of fighting, feeding, and rampaging dinosaurs. Of course, most of the "dinosaurs" in the movies were the impossibly huge beasts of the monster-runs-amuck genre, such as *Godzilla, Gorgo,* and *The Beast from 20,000 Fathoms.* Though wildly inaccurate, as every dinosaur buff knew, these movies did inspire the juvenile imagination. My 1964 Aurora Plastics Corporation Godzilla model remains a proud possession.

Children's books, like *The Big Golden Book of Dinosaurs,* provided beautiful illustrations of dinosaurs and information that more or less accurately reflected the scientific views of the day. The books gave confident answers about what dinosaurs were and how they lived: Dinosaurs were big, cold-blooded reptiles—fierce but lumbering and dimwitted. The emphasis was on size, power, and stupidity.

The archetypical dinosaur was *"Brontosaurus."* The books I read told me that *Brontosaurus* meant "thunder lizard" since the earth must have thundered when it walked. But the books usually depicted *Brontosaurus* as chest-deep in water, lazily munching succulent plants. They stayed in the water both because it supported their bulk and to escape *Allosaurus,* which for some reason could not swim. It is easy to see how the torpid, swamp-bound dinosaur became a metaphor for obsolescence.

The 1970s brought a revisionist spirit to dinosaur studies. People talked about a "dinosaur renaissance," or even a "revolution" in paleontology. Mavericks like John Ostrom and (more so) Robert Bakker promoted a new image of dinosaurs: agile, energetic, maybe even warm-blooded—more like super birds rather than lethargic mega-lizards. Bakker's dinosaurs were lean and mean; his *Tyrannosaurus* sprinted at 40 m.p.h. and *Triceratops* galloped rather than lumbered. Even the huge sauropods were portrayed as active dry-land dwellers rather than swamp-bound sluggards. Bakker's claims, particularly his assertions about dinosaur physiology, were challenged by more conservative paleontologists. The resulting controversy quickly overheated and spilled over into the popular media.

A much nastier dispute erupted in the early 1980s. The cause of the sudden extinction of the dinosaurs had long been a subject of earnest, and occasionally fatuous, speculation. The most respectable conjectures postulated a gradual climatic cooling and drying combined with increased competition from the ascendant mammals. Then, in June 1980, Nobel Prize–winning physicist Luis Alvarez and coauthors published an article in the journal *Science* that attributed the extinctions at the end of the Cretaceous to the impact of a 10 km diameter asteroid and the resulting worldwide deep freeze caused by the blockage of sunlight by impact ejecta. This radical hypothesis, so opposed to the gradualistic causes favored in the earth sciences, soon provoked a series of furious cross-disciplinary controversies.

The controversies of the 1970s and 1980s left many questions about dinosaurs unresolved. Were they warm-blooded? Did birds descend from dinosaurs, or did both evolve from earlier reptiles? Were dinosaurs good parents, nurturing their young in nests, as paleontologist John Horner suggests? Was *Tyrannosaurus* a predator, as has always been assumed, or a scavenger? Could it run at 40 m.p.h. as Bakker claimed? Did an asteroid impact eradicate the dinosaurs, or was the extinction more gradual and mostly due to earthly causes? The layperson, reviewing these controversies, will wonder how much is really known about dinosaurs.

Paleontology is an inexact science and has recently suffered more than its share of imbroglios. Disputes between scientists, especially when rancorous, protracted, and public, often promote cynicism about science. If scientists possess reliable methods for acquiring knowledge, why are their disagreements so inconclusive? One of the aims of this book is to guide readers through the complexities of some of these debates and to show that there is actually more consensus than often appears.

A deeper reason for increasing skepticism about science is that science is under strident ideological attack from both the right and the left. From the right, "scientific creationists" and proponents of "intelligent design theory" attack geology, paleontology, and evolutionary biology as well as the naturalistic presuppositions of science. They ar-

## Introduction

gue that much of current science is a front for dogmatic materialism and naturalism. Activists such as Phillip Johnson, and even a few scientists such as Michael Behe, call for the demise of Darwinism and the resurrection of theistic science. I am sad to say that a few philosophers whom I had previously held in deepest respect have joined the creationist camp (see Pennock 1999, Miller 1999, and Eldredge 2000 for devastating critiques of the "new" creationism).

From the left, a variety of postmodernists, Marxists, feminists, literary critics, radical ecologists, sociologists, and others have sought to debunk the traditional image of science as objective knowledge. They charge that the guiding values and methods of science are pervaded with reactionary, environmentally destructive, and patriarchal assumptions. Some of their more vaporous musings are hard to understand clearly, for instance, the charge that science adheres to a Western, linear, masculine (all bad things) way of thinking or presupposes the "metaphysics of presence." However opaque their rhetoric, their aim is clear enough: They want to deflate science, to cut it down to size, and to display it as no more "rational" or "objective" than any other form of discourse.

As one whose education began in the immediate post-Sputnik era, these ideological assaults on science seemed odd and disturbing to me, especially the ones from the left. I had been taught that science was good, a force for progress and enlightenment and the most effective foil for obscurantism. From growing up in the Deep South, I knew all about fundamentalist antipathy to science; sweaty evangelists fulminating against godless "evil-lution" were nothing new. The defection of many leftist scholars to trendy schools of antiscience was an unexpected betrayal.

As I studied the sociology, philosophy, and history of science, I found that the roots of skepticism about science run deep. Such skepticism is not simply an effusion of the sillier ecofeminists or literary critics (though some very silly stuff has gained currency in academe). Serious scholars have raised serious issues which must be addressed in detail.

Scientists and their philosophical defenders have rallied against the left-wing assaults with countercritiques of equal astringency. A potent counterattack was the 1994 book *Higher Superstition: The Academic Left and Its Quarrels with Science* coauthored by Paul R. Gross, a biologist, and Norman Levitt, a mathematician. This was followed in 1996 by the anthology *The Flight from Science and Reason,* edited by Gross, Levitt, and Martin W. Lewis. A more recent riposte is *Fashionable Nonsense: Postmodern Intellectuals' Abuse of Science* (1998) by Alan Sokal and Jean Bricmont.

The often acrimonious exchanges between the science critics and their critics have come to be called the "science wars." It is easy to see why the rhetorical temperature of these debates has been high. The two sides are polarized, and what is at stake is the status of scientific knowledge in our culture. Various labels have been used to characterize the two sides in the science wars: objectivists versus relativists, realists versus antirealists, traditionalists versus postmodernists. In chapter 4 I portray the science wars as a conflict between rationalists and constructivists. These terms are defined carefully there. Informally, rationalists believe that scientific consensus is ultimately driven by nature; constructivists say that it is driven by society. Rationalists think that scientists *discover* truths about the natural world; constructivists hold that scientists *construct* those "truths."

So are dinosaurs social constructs? Do we really *know* anything about dinosaurs? Might not all of our beliefs about dinosaurs merely be figments of the paleontological imagination? A few years ago such questions would have seemed preposterous, even nonsensical. Now they must have a serious answer.

Of course, in a certain sense, it is a trivial truth that all knowledge is a social construct. Theories do not grow on trees, and hypotheses do not fall from the sky like rain; scientific knowledge is a product of scientific communities. However, when postmodernists, sociologists of knowledge, and others now characterize science as a social construct, they mean much more than this. They mean that science is a kind of sophisticated mythology, the folk beliefs of the tribe of scientists. As culturally constructed artifacts, scientific beliefs are no more or less true than Zande beliefs in witches or Australian Aborigines' stories of the "dream time." In short, social constructivism is a radically skeptical and relativist view of science.

The question of how much we really know about dinosaurs can therefore only be answered in a wider context that addresses skepticism about science in general. This is what I pursue in this work: Case studies of important controversies in dinosaur paleontology provide the basis for examining skeptical claims about science. By confronting skeptical arguments in this way, I hope to rebut those arguments and to justify continued confidence in the procedures and results of science in general and of paleontology in particular.

I argue that our beliefs about dinosaurs are not entirely social constructs, that is, that the global relativism, skepticism, and cynicism of the constructivists is not warranted. On the contrary, many of our paleontological beliefs are well constrained by physical reality, a reality which is not socially constructed and is accessible by reliable scientific methods. I am not denying that social factors play a fundamental role in the practice of science. Science *is* a social activity, a collective knowledge-making enterprise and therefore as much subject to sociological analysis as any other collective human endeavor.

What I oppose is the reductionist view that social factors are the *whole* story, and that traditionally conceived rational and evidential (i.e., epistemic) factors play no crucial role in forming scientific beliefs. I also deny that truth is a matter of convention and that the methods of science are mere "rules of the game" established by custom or historical accident. I hold both that objective rational standards exist, and that scientists actually employ such standards in their theory-choice deliberations.

This book contains considerable polemical content and so (alas!) will be another shot fired in the science wars. Rational people prefer cooperation to combat, but sometimes there is no alternative but to fight. Despite some hopeful signs I note in the book, the battle has yet to be won. Rationalists have a duty to win. The reason is that relativism, constructivism, and other forms of epistemological subjectivism are false (or incoherent) and pernicious. Thomas Nagel eloquently expresses the danger:

> The worst of it is that subjectivism is not just an inconsequential intellectual flourish or badge of theoretical chic. It is used to deflect argument, or to belittle the pretensions of the arguments of others. Claims that something is without relativistic qualification true or false, right or wrong, good or bad, risk being derided as expressions of a parochial perspective or form of life—not as a preliminary to showing that they are mistaken whereas something else is right, but as a way of showing that nothing is right and that instead we are all expressing our personal or cultural points of view. The actual result has been a growth in the already extreme intellectual laziness of contemporary culture and the collapse of serious argument through-

out the lower reaches of the humanities and social sciences, together with a refusal to take seriously, as anything other than first-person avowals, the objective arguments of others. (Nagel 1997, pp. 5–6)

Still, in the final chapter I suggest how some rapprochement might be possible. It may seem odd or even hypocritical to offer an olive branch after vigorously wielding the mace, but a warrior may seek an honorable peace after the most implacable foes are vanquished.

The first three chapters of the book present three case studies from the history of dinosaur paleontology. Chapter 1, "Mr. Carnegie's Sauropods," details a particularly interesting episode: For forty-five years the Carnegie Museum in Pittsburgh, one of the world's great natural history museums, displayed one of its prize specimens with the wrong head. During that time *Apatosaurus louisae,* a showpiece of the museum's Dinosaur Hall, was displayed with what is now identified as a *Camarasaurus* skull. *Camarasaurus* and *Apatosaurus* are two very different sorts of sauropods. More importantly, the vertebrate paleontology community accepted this amalgamation as the true *Apatosaurus.*

Prima facie there is much grist here for the constructivists' mill. Was not the *Apatosaurus* accepted for forty-five years a social construct? If so, how many currently accepted "reconstructions" might actually be such constructs? I trace the convolutions of this story, making every effort to present it "warts and all." The story I uncover is complex and resists reduction to *any* stereotype of scientific practice.

Chapter 2, "The Heresies of Dr. Bakker," deals with a more recent and better publicized controversy—the dispute over Robert Bakker's claim that dinosaurs were "warm-blooded." Starting in the late 1960s, Bakker, the *enfant terrible* of dinosaur paleontology, became a vocal advocate of the view that dinosaurs were endotherms and concomitantly had very active lifestyles. This contention contradicted accepted opinion and precipitated a lively controversy. The dispute peaked at a 1978 American Association for the Advancement of Science symposium on Bakker's theories.

I trace the development of Bakker's argument from his first publication in 1968 to the full development of his case in the mid-1970s. The early criticisms of Bakker and his replies are also examined. Finally, the salient criticisms from the 1978 AAAS symposium are presented in some detail. The verdict rendered by the symposium participants (except, of course, Bakker) was that the hypothesis of dinosaur endothermy remained "not proven."

Chapter 3 takes up "the 'conversion' of David Raup." When scientists reject a reigning theory and accept a new one, do they undergo something like a religious conversion? Does their new outlook change their whole perspective on reality and radically reorient their approach to their field? Talk of "conversions" and "gestalt switches" among scientists became popular following the publication of Thomas Kuhn's very influential book *The Structure of Scientific Revolutions*.

If Thomas Kuhn had not written *The Structure of Scientific Revolutions*, social constructivism might never have existed. It was Kuhn's book that inspired sociologists to snatch science studies from the philosophers. Kuhn was taken as arguing that the proponents of different theories are hermetically sealed into "incommensurable" worldviews, that is, that their perspectives are so radically discrepant that they cannot even meaningfully disagree. It follows that scientists cannot compare competing theories in the light of neutral choice-criteria. There are no such neutral criteria; each ruling theory, or "paradigm," incorporates its own exclusive set of standards. Most obnoxious to philosophers was the implication that these different theories are so insulated that their supporters are described as living in different "worlds" and are capable of changing theories only by a process of semireligious "conversion."

Noted paleontologist David Raup claims in his book *The Nemesis Affair* to have undergone a conversion-like experience. Here he reports his reaction to the Alvarez asteroid-extinction claims and the development of his own extinction hypothesis. He portrays himself, like the Apostle Paul, as one who originally persecuted the new doctrine but who soon saw the light and became an enthusiastic convert.

Debate over the interpretation of Kuhn continues to this day. It would be useful to follow a particular scientist through a process of "conversion" from one "paradigm" to another, and Raup's *The Nemesis Affair* seems tailor-made for that purpose. In chapter 3 I carefully examine some of Raup's work before, during, and after his alleged transformation and find no evidence of a radical methodological or conceptual hiatus in his approach to science. In other words, I find no evidence that he underwent a "world-change" or semireligious "conversion."

## Introduction

Chapter 4, "Are Dinosaurs Social Constructs?" begins my critical analysis of current skepticism about science. Social constructivism is closely associated with the burgeoning fields of "science and technology studies" and the "sociology of scientific knowledge." These are big fields with many practitioners. Further, the scholars in these fields have often changed their views over the twenty or so years since work in these areas really got going. A systematic survey of constructivist opinion would therefore be impossible (see Golinski 1998 for a sympathetic overview). I shall therefore select one constructivist as representative and shall focus on three of his works written over a nine-year period.

Throughout his career, sociologist Bruno Latour has typified the spirit of the postmodernist or constructivist science critic (though he vehemently repudiates both the "postmodernist" and "constructivist" labels). Irreverent, provocative, even outrageous, his oracular style and protean opinions make him the perennial gadfly of science studies. His 1979 book *Laboratory Life* (written with Steve Woolgar) was subtitled "The Construction of Scientific Facts" and pioneered the social constructivist view of science. Latour and Woolgar claimed that scientific "facts" are simply the stories that emerge when scientists reach consensus. "Reality" and "nature" are therefore whatever scientific convention defines them to be. Worse, in negotiating consensus scientists employ rhetoric that disguises the constructed nature of hypotheses and deludes them into thinking that they are discussing "objective" matters.

In his later books, written in the mid-1980s, Latour took a different slant, depicting the scientific process as a sort of Hobbesian war of all against all. In this scientific free-for-all, the chief weapon is rhetoric, which is used to beat down opponents. I refer to the case studies in comparing Latour's view with the very different analysis of scientific rhetoric offered by Marcello Pera in *The Discourses of Science* (1994). My conclusion is that, though these debates featured plenty of abrasive Latour-type rhetoric, logical argument and the weight of evidence, the traditional epistemic factors, still were crucial. I conclude that Latour's constructivism and cynicism are unwarranted.

In the early 1990s Latour claimed to renounce social constructivism; by 1995 any characterization of Latour as a constructivist would elicit a reaction of extreme indignation (Latour 1995). His new line of argument declares that rationalism and constructivism both err

## Introduction

by demarcating the natural from the social; they disagree only over whether nature or society drives the course of science. Latour proposes, obscurely, that we dissolve the natural-social distinction and see the entities postulated by science as "quasi objects" that can be viewed as more or less social or more or less natural on different occasions (Latour 1992, 1993).

In his highly entertaining (and exasperating) *The Last Dinosaur Book* (1998), W. J. T. Mitchell in effect adopts Latour's recommendation and provides an in-depth interpretation of dinosaurs as "quasi objects": That is, he argues that the "real" dinosaur can no longer be distinguished from the cultural icon it has become. Chapter 5 examines Mitchell's claims. Although with dinosaurs it is often very difficult to say where the "social" ends and the "natural" begins, Mitchell badly overstates his case. I argue that he commits a number of lapses of logic typical of the constructivist-postmodernist analyses, and that, despite Barney, Godzilla, and *Jurassic Park,* we can still separate dinosaur fact from dinosaur fiction.

The most serious constructivist charge is *not* that science is determined by nonrational social causes rather than objective evidence. Their most damning claim is that the scientific notions of "rationality," "objectivity," and "evidence" are *themselves* socially constructed through and through. In other words, they hold that the canons of reason in science are themselves social conventions, mere "rules of the game" contingently adopted by scientific communities.

What would lead someone to think that the methods and standards of science are just arbitrary rules of the game? I think that the study of the history of science is the chief motivation for this charge. The history of science has often been told as a tale of linear progress: Past ignorance and obscurantism steadily give way to present wisdom. When *not* viewed in such a triumphalist vein, the history of science reveals disturbing facts. We find that even the greatest scientists ground ideological axes. Also, views which in hindsight we call "progressive" were no less ideologically infected than their "obscurantist" competitors. Since the winners write the textbooks, they could enshrine their own preferred methodologies. Had the other side won, science might now be done very differently.

Telling the history of science as a march of triumph is now derided as "Whig history." For contemporary historians of science, there is no sin graver than to write Whig history, that is, history that evaluates past episodes in the light of how well they prefigure or pave the way for present science. Undoubtedly, scientists and their achievements should be judged by the standards appropriate for their times, not ours. However, an overly restrictive interpretation of the ban on Whig history has permitted excessive relativism and constructivism to enter into the practice of historians of science.

Chapter 6, "History, Whiggery, and Progress," examines a historical study of a particular episode in the history of paleontology. In his 1982 book *Ancestors and Archetypes,* historian Adrian Desmond analyzes the debate between T. H. Huxley and Richard Owen on the connection between birds and dinosaurs. Desmond argues that Huxley, whose views supported the "progressive" evolutionary theory, was just as ideologically driven as Owen.

While Desmond's historical analysis of Huxley and Owen is solid, his results should not be extrapolated too far. If science were merely an excrescence of ideology, it is a puzzle how it could progress, and science *does* progress—in ways that sociologists cannot explain away. Scientific advance crucially involves progress in method. Constructivists attempt to debunk scientific methods by calling them mere "rules of the game," that is, not objectively reasonable (and adopted for that reason) but (ultimately arbitrary) products of historical accident. I review some of the many ways that paleontological methods have changed since Huxley's and Owen's day and conclude that these methods are genuinely progressive, that their adoption by paleontologists was warranted (because they *really are* better ways of doing science), and thus that their adoption was not merely contingent. Along the way I criticize the best-known defense of the contingency of scientific methods—Steven Shapin and Simon Schaffer's *Leviathan and the Air-Pump* (1985). I also criticize the simplistic construal of the ban on Whig history that has encouraged excessively constructivist analyses in the history of science.

My concluding meditation "Beyond the Science Wars" considers possible grounds for rapprochement between the two camps. First, after saying much in criticism of the sociology of science, I need to show that it can be useful for the understanding of paleontology. I do this by suggesting a *possible* sociological analysis of the warm-blooded dinosaur debate. My aim is to show that the turbulent social conditions and counterculture of the late 1960s might have influenced Bakker and so the course of the debate. Second, and more importantly, I explore a model of science that might satisfy the intuitions of rationalists while accommodating the genuine insights of the constructivists.

This model was developed by Richard Bernstein as a reinterpretation of Kuhn (Bernstein 1983). Bernstein argues that Kuhn, rightly construed, saw science as an eminently rational process and endorsed all of the traditional theory-choice criteria recognized by rationalists. He was merely debunking the view that scientific procedure could be rigidly specified by a series of algorithms or exhaustively characterized by formal rules. Bernstein claims that Kuhn does not say that theory-choice decisions are irrational *conversions,* but are rather the result of fully rational *persuasion.* As Bernstein sees it, Kuhn's innovation was to argue that theory choice is a matter of *practical* reasoning instead of the automatic application of formal procedures. (Kuhn himself vigorously denied the imputations of irrationalism; see especially the essays in his 1977 book *The Essential Tension.*)

If that was Kuhn's main intention, it sounds much like the aim of Harry Collins, long regarded as a leading defender of constructivism. In his 1985 book *Changing Order,* Collins criticizes what he considers the canonical "algorithmic" view of science and offers an alternative "enculturational" model. Collins supports his claims with the concept of the "experimenter's regress," which has usually been taken as arguing that experimental evidence is inevitably circular. Naturally, read this way, the regress claim drew much fire.

Perhaps, however, Collins and his critics have misunderstood key points. I seriously doubt that mainstream, "canonical" philosophy of science has ever viewed theory choice as algorithmically determined or exhaustively prescribed by formal rules. On the other hand, perhaps Collins's regress argument can be interpreted as implying not inevitable circularity, but only that the replication of experiment is always complex, contextual, and never automatic. If these construals are acceptable, perhaps Collins and his critics can find common ground in the Kuhn-Bernstein model of science. If so, then, just maybe, we can see hope for the culmination of the science wars.

The works of constructivists examined here were published from fifteen to more than twenty years before this book was written. Why not examine more recent works in the rapidly evolving field of the sociology of science? Also, these works have already received much analysis and commentary. Why return to rather well-trodden ground?

A systematic survey of the sociology of knowledge would be impossible here. I chose books that I regard as the "classics" of the field; they remain highly influential and serve as benchmarks for more recent developments. Also, I am aiming at an audience wider than professionals in the history, philosophy, or sociology of science. For instance, I doubt that most practicing paleontologists have taken the time to go through the convolutions of *Laboratory Life*. I have tried to fill in the details for those who have had neither time nor inclination to read the founding documents of social constructivism.

Finally, readers will notice that the tone of this book is more personal than is usual for scholarly works, with frequent use of the first person to indicate my own reactions and opinions. The issues discussed here *are* of deep personal significance for me and many others. As I said earlier, what is at stake in the science wars is the status of scientific knowledge in our culture. Since science and technology affect all of us profoundly, these are no small stakes.

Constructivists, full of leveling zeal, love to focus on the naked emperors and clay-footed giants of science. It is no surprise that they find plenty of them. The really surprising fact, maybe the most astonishing thing in history, is that we ever had the chutzpah to claim to *know* anything at all about the inner workings of this stupefyingly complex cosmos. The world looks like what our ancestors thought it was—the abode of powerful and unpredictable supernatural forces. It is amazing that we ever came to see the universe as anything else. But we do. Even the most stringent science critics get flu shots and wash their hands before eating.

So, courageously or foolishly, we humans have embarked on the scientific enterprise. Einstein famously said that the known is infinitesimal compared to the unknown, yet the little we know is our most precious possession. Was he right? The culmination of the science wars will determine whether the next generation will be told that science is a precious possession or just another castle erected on the clouds of human fancy and folly.

# DRAWING OUT LEVIATHAN

# 1

# MR. CARNEGIE'S SAUROPODS

I BEGIN WITH two scenarios:

(1) On a spring morning in a long-vanished world, an enormous creature was dying. Driven by thirst, she trudged toward the slow, muddy waters of a great river. Her massive legs could no longer support her, and she fell into the riverbed to die. Before scavengers could tear and scatter the carcass, a violent storm washed it downstream. It came to rest on a sandbar in the middle of the meandering river.

As the body decayed, the river shrouded it with sediment. Soon only the bones were left, encased in fine sand and silt. As the ages passed, the sediment over the bones deepened and hardened, turning into rock. Molecule by molecule, the bones themselves were also changed into rock and buried deep in the crust of the earth. Through time of unimaginable duration the fossilized bones lay hidden. All of the dead creature's kin also disappeared, to be replaced by other faunas, and those with yet others again and again.

The earth itself changed, as vast tectonic forces crushed and squeezed the continents into new shapes. Eventually those forces thrust the bones and their tomb of rock up from the depths. Wind and rain did their slow erosive work until, after one million five hundred thousand centuries of burial, the bones were brought into the sunlight.

(2) On an autumn morning in New York City in the year 1898, the richest man in America had his breakfast ruined by a newspaper story. Leaving his kipper and egg untouched, he fumed over the lead article. The paper told of the discovery of the "world's most colossal animal" by a bone hunter working for New York's American Museum of Natural History. The accompanying illustration showed the creature, rearing on pillar-like legs to peer into an eleventh story window.

Andrew Carnegie was galled to think that another museum, and

# 2
## Drawing Out Leviathan

not his Carnegie Museum in Pittsburgh, would have the glory of housing the skeleton of the biggest animal. It was worse to contemplate the credit that would go to J. P. Morgan, a chief benefactor of the American Museum, and one of the very few men who could match Carnegie in a business deal. Incensed, Carnegie wrote a check for $10,000, quite a sum in those days, and sent it to W. J. Holland, director of the Carnegie Museum, with curt instructions to buy the skeleton of the colossal animal. Morgan's minions may have found the damn thing, but it would stand in *his* museum.

★ ★ ★

The Carnegie Museum of Natural History in Pittsburgh houses one of the world's great collections of dinosaur fossils. The holotype skeleton of *Tyrannosaurus rex* is there, the type specimen that defines the species. *Stegosaurus* and *Allosaurus* confront the visitor at the entrance to Dinosaur Hall. A beautiful, complete skeleton of a juvenile *Camarasaurus* is on display, still partially encased in the original matrix of rock. Two skeletons dominate the hall: the enormous sauropods *Diplodocus* and *Apatosaurus*. This chapter tells the story of the discovery and reconstruction of these two specimens, highlighting the controversy over the head of *Apatosaurus*.

No specimen of *Apatosaurus* has been found with the skull in place at the end of the neck. O. C. Marsh, the discoverer of *Apatosaurus,* reconstructed the dinosaur with a *Camarasaurus*-like skull. Such a reconstruction became the received image of *Apatosaurus* within the paleontological community (Berman and McIntosh 1978; Norman 1985). W. J. Holland, director of the Carnegie Museum, dissented. He held that a *Diplodocus*-like skull found in close conjunction with the *Apatosaurus* specimen was in fact the correct skull.

This dispute did not end quickly; the final episode (so far) was played out nearly seventy years after the discovery of the Carnegie's *Apatosaurus*. Holland chose not to attach the *Diplodocus*-like skull to the Carnegie's *Apatosaurus*. Instead, the specimen was displayed headless for nearly twenty years. In 1934, after Holland's death, a mold of a large *Camarasaurus* skull was attached to the skeleton. Finally, in 1979, following extensive studies by two noted paleontologists, David Berman and John McIntosh (recognized as the world's leading authority on sauropods), the camarasaurid skull was removed, and a mold of

# 3
# Mr. Carnegie's Sauropods

Holland's *Diplodocus*-like skull was attached (Berman and McIntosh 1978). For a period of forty-five years, the Carnegie Museum presented one of its foremost specimens with what is now regarded as the wrong head. How this happened at one of the world's leading museums of natural history will be the focus of this chapter.

Despite its importance to paleontology, this story has not received the attention it deserves from professional historians of science. An interesting but somewhat sketchy account is given in Helen McGinnis's *Carnegie's Dinosaurs* (1982). I shall reconstruct these controversies from the original published sources and from unpublished records in the archives of the Department of Vertebrate Paleontology of the Carnegie Museum of Natural History.

Sauropods are the stereotypical dinosaurs of popular imagination: huge body, columnar legs, long tail, and a long neck tapering to a relatively tiny head. Some of the best known of the dinosaurs, such as *Apatosaurus, Diplodocus,* and *Brachiosaurus,* were sauropods. The sauropods were the largest dinosaurs, and the largest terrestrial animals, some reaching forty meters in length and others weighing perhaps up to 100,000 kg (McIntosh, Brett-Surman, and Farlow 1997, p. 281). It was the stupendous size of the sauropods that fired the imagination of Andrew Carnegie.

By the end of the nineteenth century, the American Museum of Natural History (hereafter AMNH) was already emerging as the nation's leading museum of vertebrate paleontology. Much of the success of the AMNH was due to the bountiful support of such multimillionaires as J. P. Morgan (Rainger 1991). According to Carnegie's biographer Joseph Frazier Wall, Carnegie's relationship with Morgan had soured in 1885, when Morgan caused a railroad deal to go against Carnegie's interests. According to Wall, Carnegie subsequently regarded Morgan with deep suspicion (Wall 1970, p. 514).

After reading about the AMNH's coup, Carnegie demanded that Holland buy the dinosaur for the Carnegie Museum. The American Museum's specimen was not for sale—there was no such specimen. William H. Reed, chief fossil collector for the AMNH and purported discoverer of the "Most Colossal Animal," had concealed the fact that his find consisted of a single bone (McGinnis 1982, p. 15). Unaware of the meagerness of the find, Holland hired Reed and sent him into Wyoming on a mission to find a dinosaur for Mr. Carnegie.

The Carnegie Museum expedition, guided by the dubious Reed,

began unpropitiously. No fossils were found, and Reed was forced to admit the truth about his exaggerated claims. The Carnegie team persevered, and they soon were rewarded beyond all expectation. On July 4, 1899, at a Wyoming site they called Sheep Creek, they came upon the rarest of paleontological finds: the nearly intact, articulated skeleton of a gigantic sauropod. In 1901 the first description of *Diplodocus carnegii* was published in the scientific literature (McGinnis 1982, pp. 14–15).

When mounted, the skeleton of *Diplodocus* was over twenty-five meters long. For many years it was the longest dinosaur ever discovered. The specimen was so large that the museum's exhibit hall had to be expanded to house it. The popular press made much of "Uncle Andy" and "Dippy." By 1912 the *Pittsburgh Gazette Times* could characterize the Carnegie Museum as a "*Diplodocus* factory" and "one of Pittsburgh's busiest industries" as requests for casts of the skeleton poured in from around the world (Anonymous 1912). Requests came from Edward VII of England, the emperors of Germany and Austria-Hungary, the president of France, and the king of Italy (McGinnis 1982, p. 17).

Basking in the attention of kings and presidents, Carnegie long remained enthusiastic about dinosaurs, particularly large sauropods. According to Elizabeth Hill, archivist for the Carnegie Museum's Department of Vertebrate Paleontology, Carnegie supported the museum's paleontological expeditions until 1915, just four years before his death (personal communication 1994). The worldwide dissemination of *Diplodocus carnegii* casts was a triumph for the Carnegie Museum and its illustrious benefactor. The AMNH and its backers could boast no greater accomplishment.

In the meantime, the museum's bone hunters had discovered a particularly rich fossil site in northeastern Utah. The Carnegie Quarry, now part of Dinosaur National Monument, was one of the most bountiful sources of dinosaur fossils ever discovered. Holland placed Earl Douglass in charge of the excavations at the quarry. Douglass's letters and reports to Holland are remarkable documents. They record triumphant discovery and bitter disappointment alongside prosaic worries about supplies, bookkeeping, and enduring winter in the wilds of Utah. Douglass's careful observations and inferences reveal an acute mind and well-trained eye. Occasional passages of self-reflection disclose a sensitive and introspective personality.

## Mr. Carnegie's Sauropods

Of course, it is the record of paleontological discovery that is most important here. The quarry revealed its riches right away. On August 23, 1909, in a letter to Douglas Stewart, assistant director of the Carnegie Museum, Douglass reported the discovery of a *Brontosaurus* (= *Apatosaurus*—this confusion of names is explained below) in an apparently perfect condition. Three days later he reported in greater detail to Holland:

> I have found what bids fair now to be an almost complete skeleton of a Brontosaurus (= *Apatosaurus*). I never saw anything that, on the surface, and so far as we have gone, had such promises of a whole thing. . . . We have exposed parts of the bones to the third or fourth vertebra anterior to the sacrum, and the bones are exactly in place, apparently except that the femur appears to have gone down several inches out of its place in the socket. I confidently expect at least to get all of the pelvic girdle, and the two hind limbs. . . . I fully expect to find all the dorsals. Of course, currents may have disturbed the thing before we get to the head but there is no indication as yet that any disturbance has taken place. (Douglass to Holland, August 26, 1909)

A fully articulated skeleton, particularly one with a skull, would be a find of the highest significance. To see why we must review the first *Apatosaurus* discoveries.

The scientific study of *Apatosaurus* began in a very confused and confusing way (Norman 1985, p. 81). Much of the confusion arose out of the famous feud between O. C. Marsh and Edward Drinker Cope. Marsh and Cope, perhaps the two most prominent dinosaur paleontologists of the nineteenth century, engaged in an acrimonious rivalry fueled by intense personal animosities (for a complete and very readable account of this feud, see Lanham 1973). Each man was eager to name more dinosaur species than his rival. This unbridled competition led to sloppiness and crucial misjudgments as descriptions were rushed into print on the basis of inadequate portions of the type skeletons (Berman and McIntosh 1978).

Marsh coined the name *Apatosaurus* and first described its fossils in 1877 (Marsh 1877). Two years later, at the Como Bluff quarry in Wyoming, Marsh's collectors discovered the two skeletons of what was believed to be another type of large sauropod (Norman 1985; Marsh 1879). This allegedly new sort of sauropod was named "*Bron-*

Earl Douglass. Photograph taken at the time he entered the service of the Carnegie Museum. Courtesy, Carnegie Museum of Natural History, Pittsburgh, Penn.

*tosaurus.*" However, *Brontosaurus* soon came to be regarded as the same type of dinosaur as *Apatosaurus* (Riggs 1903; McGinnis 1982, p. 71). Under the rules of paleontological nomenclature the correct name for the genus is therefore *"Apatosaurus."* However, the name *Brontosaurus* (thunder lizard) has always had more cachet than *Apatosaurus* (deceptive lizard), so *Brontosaurus* was the name that stuck (though some

ciety and his refusal to mount the skull chosen by Marsh on *Apatosaurus louisae* was already sufficient provocation, should anyone care to take offense.

Rather than timidity, one explanation of Holland's decision seems to be that he carefully discriminated between the evidence against Marsh's restoration and the evidence in favor of his own. He was much more convinced that Marsh's association was wrong than that his was correct. In the address challenging Marsh, Holland admitted that unquestionable specimens of *Diplodocus* had been found in the same layer as the skull he attributed to *Apatosaurus* (Holland 1915, p. 274). He therefore recognized that the skull could have belonged to a *Diplodocus*, but admitting this did not compromise his criticism of Marsh. More importantly, as long as Douglass's excavations at the Carnegie Quarry continued, as they did for some years, there was always the chance that a skeleton would turn up with the skull still in place. Holland specifically mentions this as his reason for not attaching a skull to *A. louisae* (Holland 1916, p. 310).

Other, more subtle social factors may also have influenced Holland. By 1915, when the skeleton of *A. louisae* was mounted, Andrew Carnegie had ceased funding the museum's paleontological expeditions, and, greatly preoccupied with world affairs, he seemed to have had few thoughts to spare for his museum (Wall 1989). Had Carnegie retained an active interest in sauropods, Holland might have felt under greater pressure to mount some type of head. A large sauropod displayed without a head, especially one named after his wife, might not have pleased Carnegie.

A skull still articulated to the neck was never found, and *Apatosaurus* remained headless when Holland died in 1932. In 1934 a mold of a camarasaurid head was mounted on the skeleton of *A. louisae*. No record remains of precisely who made the decision to attach this head or why that person did so. The documents relating to this incident are frustratingly vague. McGinnis's history says only that a head "was mounted" (McGinnis 1982, p. 73). David Berman and John McIntosh's careful historical review names no one as responsible for the mounting (Berman and McIntosh 1978). Paleontologist C. W. Gilmore, who published an authoritative study of the osteology of *A. louisae* in 1936, would say only that "it was decided" to mount a replica of a camarasaurid skull (Gilmore 1936, pp. 189–90). Andrey Avinoff, director of the Carnegie Museum in 1934, in his monthly report for

## 12
## Drawing Out Leviathan

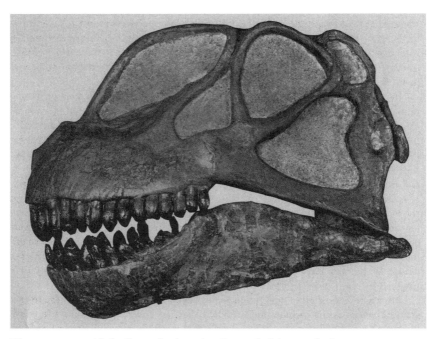

The camarasaurid skull attached to the Carnegie Museum's *Apatosaurus louisae*. Courtesy, Carnegie Museum of Natural History, Pittsburgh, Penn.

May of that year says that the replica of the skull "was mounted" on the skeleton and fails to mention whose decision it was (Avinoff 1934). Carnegie Museum archivist Elizabeth Hill informs me that she has never seen a document naming a particular individual as responsible (personal communication).

Whoever made the decision, whether it was an individual or, more likely, a committee, it is possible to infer a number of factors that likely went into the decision. By 1934 the opinion of most dinosaur paleontologists was that *Apatosaurus* was more closely related to *Camarasaurus* than to *Diplodocus* (Berman and McIntosh 1978, pp. 11–12). Thus, the prevailing scientific view did not oppose the mounting of a *Camarasaurus*-like skull on *A. louisae,* and in fact rather favored it.

Berman and McIntosh argue that the view that *Camarasaurus* and *Apatosaurus* were closely related was itself based on a series of mistakes and misjudgments (pp. 7–12). This is an interesting subplot that is

# 13
# Mr. Carnegie's Sauropods

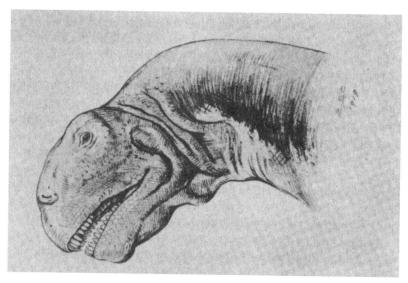

Reconstruction of the head of Cope's *Camarasaurus*. Illustration by H. F. Osborn and C. C. Mook, 1919.

worth recounting in some detail. First, not only was Marsh's reconstruction given a *Camarasaurus* skull, but other features of that dinosaur were incorporated. For instance, the ulna and manus of the *Brontosaurus* depicted by Marsh had in fact come from the partial skeleton of a large *Camarasaurus*. Also, as Holland realized, and Marsh did not, *Brontosaurus* possessed a long "whiplash" tail like *Diplodocus* and unlike *Camarasaurus* (p. 7; Holland 1915).

Marsh was not incompetent. Berman and McIntosh argue that his mistakes were due to his feud with Cope, which was at its height at the time of the *Brontosaurus* and *Camarasaurus* discoveries. They charge that this feud led to sloppiness and poor judgment as descriptions were rushed into print on the basis of inadequate portions of the type skeletons (Berman and McIntosh 1978, p. 11).

One particular source of confusion was that the half dozen or so partial skeletons that Marsh had definitely identified as *Camarasaurus* were subadult specimens, much smaller than the two large *Brontosaurus* skeletons he possessed. Berman and McIntosh claim that this led him to misidentify as *Brontosaurus* the few adult *Camarasaurus* specimens he possessed. They speculate that had Marsh realized that *Camarasaurus* grew as large as *Brontosaurus,* he might not have used adult *Camarasaurus* elements in his restoration (ibid.).

Remarkably, Marsh had within his possession, from the time of his earliest sauropod discoveries, good evidence of a close association between an *Apatosaurus* skeleton and elements of a *Diplodocus*-like skull (pp. 6–7). The quarry in which Marsh's men had discovered the first *Apatosaurus* fossils also contained a large femur. Marsh decided that the femur belonged to a new species of sauropod which he named *Atlantosaurus immanis.* However, S. W. Williston, Marsh's assistant, expressed the opinion to Marsh that *Atlantosaurus* and *Apatosaurus* were actually the same species. Further, Williston identified as belonging to *Atlantosaurus* part of a cranium which he believed to have come from the same quarry (there was some confusion over this point due to sloppiness in cataloging). This cranium had decidedly *Diplodocus*-like features.

Marsh, however, identified the cranium as belonging to a different species that had been found in a different quarry (Marsh 1896). He therefore did not associate the *Diplodocus*-like cranium with the quarry in which his *Apatosaurus* specimens had been found. Had he regarded the cranium as coming from the same quarry as *Apatosaurus,* and had he recognized its *Diplodocus*-like features, this might well have influenced his judgment about sauropod heads.

The upshot is that by 1934, despite the availability of accurate reconstructions of the postcranial anatomy of *Apatosaurus, Camarasaurus,* and *Diplodocus,* the idea of a close affinity between *Apatosaurus* and *Camarasaurus* was well entrenched (Berman and McIntosh 1978, p. 11). The decision to affix the *Camarasaurus*-like skull was therefore consistent with prevailing scientific opinion.

More importantly, in 1934 C. W. Gilmore, noted paleontologist with the United States National Museum, visited the Carnegie in order to prepare an extensive monograph on the osteology of *Apatosaurus.* Apparently, Gilmore's visit was the impetus for finally affixing a head to the skeleton. Andrey Avinoff, director of the museum, noted Gilmore's visit in his monthly report for May and in the same paragraph reported the mounting of the head: "The head of the reptile [*Apatosaurus*], found at the time near the huge skeleton of the Apatosaurus in the rocks, was identified with the best available probability as belonging to this animal, and was placed on exhibition, whereas

## 15
## Mr. Carnegie's Sauropods

the replica in plaster of Paris was mounted on the skeleton of the huge fossil which has stood headless during all these years" (Avinoff 1934). This passage indicates that the Carnegie staff had come to regard the *Camarasaurus*-like skull as the one that had been found in close association with the *Apatosaurus* remains.

Gilmore did not accept such a complete about-face, but he did regard Holland's view as wrong, as he explained in his monograph. Gilmore rebutted Holland's claim, made in the 1915 paper, that the *Diplodocus*-like skull, CM 11162, was the one found near the *Apatosaurus* skeleton:

> In this he [Holland] was in error for Mr. J. LeRoy Kay, who was Douglass's assistant at the time these skulls were collected and who was largely responsible for the preparation of the quarry map... informs me that this large Diplodocid-like skull, No. 11162 C.M., came from the western end of the quarry nearly 100 feet distant from the nearest bones of the *Apatosaurus* skeleton and in the lowermost part of the bone-bearing stratum. Likewise the second skull, No. 12020 [the *Camarasaurus*-like skull, the cast of which was mounted on the *Apatosaurus*], discussed by Doctor Holland was found somewhat farther away, but higher up in the fossil bearing strata at practically the same level as the *Apatosaurus* skeleton. These observations by Mr. Kay have been verified from the original records, so that the question of near proximity as an argument for the association of either of these skulls with the skeleton no longer obtains. The undoubted Diplodocid affinities of skull No. 11162 C.M. though larger than any *Diplodocus* cranium previously known, is sufficient in my estimation to exclude it from further consideration in this connection. (Gilmore 1936, p. 188)

In other words, Gilmore believed that *neither* skull had been found in close association with the skeleton. He regarded CM 11162 as definitely ruled out, but he did not endorse the head actually mounted and regarded its mounting as merely a "temporary expedient" until further discoveries should reveal the true nature of the *Apatosaurus* skull (p. 190). The "temporary expedient" was to remain in place for the next forty-five years. During that time both the scientific and the popular press invariably depicted *Brontosaurus* with a *Camarasaurus*-like head. By the mid-1950s even Alfred Sherwood Romer's authoritative *Osteology of the Reptiles* would depict *Brontosaurus* with such a skull (Romer 1956, p. 146).

# Drawing Out Leviathan

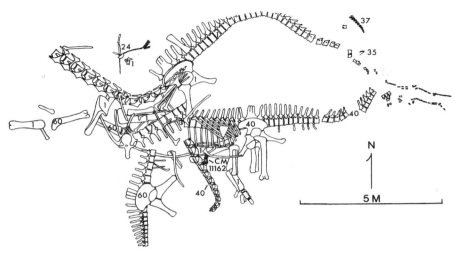

Quarry map of *Apatosaurus louisae*, showing position of the skull designated CM 11162. Courtesy, Carnegie Museum of Natural History, Pittsburgh, Penn.

In sum, the scientific evidence in 1934 somewhat supported the decision to mount a camarasaurid head on *A. louisae*. Tradition, going back to Marsh, supported such a mounting, and paleontologists had come to accept a fairly close affinity between *Camarasaurus* and *Brontosaurus*. However, Gilmore could find no sufficient reason to endorse the mounting. It therefore seems unlikely that the decision to mount the head was based exclusively upon scientific considerations. Supposing that it was not motivated solely by scientific judgment, what else may have motivated the decision?

The Carnegie Museum was not merely a research institution. Another major function of a museum is to entertain and educate the public. In theory, the relation between these two roles is supposed to be unilateral, with research influencing the presentation of science to the public, but the exigencies of public education not influencing scientific decisions. In practice, the two roles are much harder to keep separate. Ronald Rainger, in his excellent study of Henry Fairfield Osborn and the AMNH, details the extent to which these two different roles overlapped at that institution (Rainger 1991).

No doubt it was a persistent source of embarrassment for some that the museum's most formidable fossil was on public display minus a head. Unlike the Venus de Milo, *Apatosaurus* did not gain charm by being displayed with parts missing. Osborn was probably expressing a rather common view among museum directors when he said that natural history displays should seek to engender a sense of awe in the spectators (Rainger 1991, pp. 119–20). A decapitated *Apatosaurus* seems much less likely to elicit the desired response.

The camarasaurid skull, on the other hand, was most impressive. *Camarasaurus* was a robust creature with a large, heavy head. The teeth were large and spatulate—clearly capable of grabbing and biting. The *Diplodocus*-like skull, on the other hand, barely larger than a horse's skull, looked unnaturally small for so massive a creature as *Apatosaurus*. The small head and feeble, peglike dentition seemed inadequate for the consumption of food sufficient to sustain *Apatosaurus*. It was not recognized until much later that *Apatosaurus* ground its food with gastroliths rather than its teeth.

Finally, Holland died two years prior to the mounting. Holland had been the main opponent of the mounting of a camarasaurid head, and his opinions were repudiated by Gilmore. In the absence of any salient scientific objections to the mounting, and given the opportunity to remedy a perennial embarrassment, a move to attach the *Camarasaurus* skull is hardly surprising. Motivations that no one would regard as strictly scientific are rather obvious in this case.

Eventually, the topic of *Apatosaurus*'s cranium piqued the interest of preeminent sauropod authority John S. McIntosh. He visited the Carnegie Museum and read the Douglass correspondence, which clearly reported the discovery of skull CM 11162 in close association with a *Brontosaurus* skeleton. Further investigation revealed that a cataloging error had misled J. L. Kay, Douglass's assistant at the excavation, when he told Gilmore that the *Diplodocus*-like skull had not been found with the *Brontosaurus* fossils (McGinnis 1982, p. 73). McIntosh concluded that Holland had been right all along about the close association of CM 11162 with the *Brontosaurus* skeleton. In 1978 McIntosh teamed up with David S. Berman, a morphologist and Assistant Curator of Vertebrate Fossils at the Carnegie, to produce the authoritative monograph that was to settle the issue of *Apatosaurus*'s head: *Skull and Relationships of the Upper Jurassic Sauropod Apatosaurus (Reptilia/Saurischia)*.

The monograph begins with a detailed and extensive "historical review" in which the authors conclude that CM 11162 was in fact the skull found in close association with the *Apatosaurus* remains. (Much

## Drawing Out Leviathan

of this evidence has been mentioned earlier in the chapter.) Most importantly, they give a point-by-point account of how, starting with Marsh, *Apatosaurus* came to be thought to possess many *Camarasaurus*-like features (Berman and McIntosh 1978, pp. 6–12). They also provide meticulous documentation in support of their claim that cataloging errors and sloppiness in record keeping had compounded confusion and inhibited clear judgment.

The monograph next provides complete descriptions of the skull of *Diplodocus* and the proposed *Apatosaurus* skull, CM 11162, and makes a careful comparison between the two (pp. 12–30). There is no need to give an extended account of the extremely technical and detailed anatomical descriptions. It is sufficient to quote a brief passage showing how Berman and McIntosh argue for the close similarity between the two skulls while noting and accounting for the differences:

> Though the incomplete preservation of the skull CM 11162 eliminates many opportunities for detailed comparisons, this skull is obviously very close to that of *Diplodocus*. Comparison between CM 11162 and *Diplodocus* skull CM 3425 and those skulls very likely to belong to *Diplodocus* have revealed a number of subtle proportional and structural differences. Some of these differences, however, have to be evaluated with caution because they may be the result of postmortem distortion of CM 11162. It will also be noticed that the obviously greater general robustness of CM 11162 is a fundamental aspect of many of the features used . . . to contrast it with the skull of *Diplodocus*. (p. 25)

For instance, the palate of CM 11162, with the exception of proportional differences (appropriate for a much more massive animal), does not noticeably depart from that ascribed to *Diplodocus* (p. 24). *Diplodocus* skulls have been found still articulated with the neck vertebrae.

Having established the close similarity between the skulls, Berman and McIntosh return to the issue of the evidence associating *Apatosaurus* remains with *Diplodocus*-like skulls. They note that, as mentioned previously, the quarry in which Marsh made his original *Brontosaurus* discoveries yielded a nearly complete cranium of a clearly diplodocid type. Also found, and likewise ignored by Marsh, were a pair of quadrates that are near duplicates of those of *Diplodocus* and

# Mr. Carnegie's Sauropods

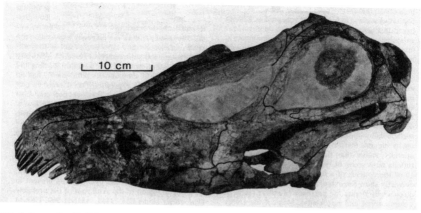

*Diplodocus* skull CM 11161. Note the similarity to CM 11162 in figure below. Courtesy, Carnegie Museum of Natural History, Pittsburgh, Penn.

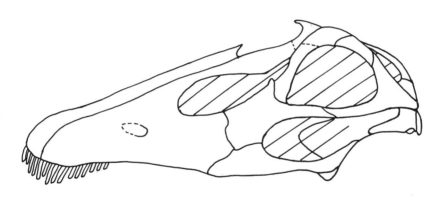

Restoration of probable *Apatosaurus* skull CM 11162. Courtesy, Carnegie Museum of Natural History, Pittsburgh, Penn.

of CM 11162 while very different from the quadrates of *Camarasaurus* (p. 27).

Finally, a close comparison is made between the postcranial skeletons of *Apatosaurus, Diplodocus,* and *Camarasaurus* (pp. 30–32). They summarize their results as follows:

> The postcranial skeletons of *Apatosaurus* and *Camarasaurus* have been generally considered more similar to each other than to *Diplodocus* because of their much greater robustness. It is in fact difficult

> to distinguish between isolated hindlimb bones of *Apatosaurus* and *Camarasaurus,* especially if these elements are imperfectly preserved. Excepting this superficial resemblance between *Camarasaurus* and *Apatosaurus,* the postcranial skeletons of *Apatosaurus* and *Diplodocus* share a large number of characters that set them widely apart from *Camarasaurus. Diplodocus* and *Apatosaurus,* in contrast to *Camarasaurus,* have relatively very long necks, short trunks, and very long tails, unusual anterior caudal vertebrae and midcaudal chevrons, shorter forelimbs and metacarpals, and reduced number of carpal and tarsal elements. (p. 30)

In short, except for differences due to its much greater robustness, the postcranial anatomy of *Apatosaurus* is very similar to that of *Diplodocus.*

Putting all the pieces of evidence together—the close association of CM 11162 with the *Apatosaurus* remains in the Carnegie Quarry, the clearly diplodocid nature of CM 11162, the record of errors that led paleontologists to associate *Camarasaurus* with *Apatosaurus,* the discovery of *Diplodocus*-like skull elements with *Apatosaurus* remains in Marsh's quarry, and the great affinity between the postcranial anatomies of *Apatosaurus* and *Diplodocus*—Berman and McIntosh became convinced that a diplodocid rather than a camarasaurid skull probably belonged to *Apatosaurus louisae.* The cumulative weight of their arguments makes quite a strong case.

The Carnegie Museum agreed with Berman and McIntosh, and on October 20, 1979, a ceremony was held and the *Camarasaurus* skull was at last replaced by a plaster cast of CM 11162. CM 11162 itself, too precious to be mounted, lies in a glass case adjacent to the skeleton. Berman and McIntosh's claim was quickly accepted by other paleontologists. By 1985, David Norman, in his highly regarded *Illustrated Encyclopedia of Dinosaurs,* simply classified *Apatosaurus* among the diplodocids (Norman 1985, p. 81).

In conclusion, we have seen that the controversies and events recounted in this chapter admit of a number of diverse explanations and interpretations. Detailed and often highly technical arguments were offered by the disputants; these have been closely examined. However, we have also seen that a number of different social factors could very plausibly be adduced to explain the course and outcome of the sauropod controversies. On the one hand, the very fact that sauropods became the center of so much attention may be attributed to the enthusiasms, rivalries, and deep pockets of millionaires such as

## Mr. Carnegie's Sauropods

Andrew Carnegie. On the other hand, the waning of Carnegie's interest in sauropods by 1915 may have meant that Holland was under less pressure to mount some sort of head on *Apatosaurus*.

The fact that museums such as the Carnegie are both centers of research and institutions of public education means that these two roles can overlap. The decision to mount a *Camarasaurus* head on *Apatosaurus louisae* may very well have been much less of a scientific decision than a public relations move. A headless sauropod is a much less awesome sight than is one with a formidable cranium.

What lessons do we draw about the nature of science from the infamous wrongheaded dinosaur incident? It is clear that science does not always proceed in the traditionally rational manner. Social influences certainly play a role, a bigger role than most scientists might be happy to admit. Nevertheless, reason and evidence, the traditionally "scientific" factors, also molded and shaped every step of our story. The desire to impress the public and end the embarrassment of a headless *Apatosaurus* may have motivated the decision in 1934 to mount the *Camarasaurus* head. However, at the time the weight of scientific opinion, including that of leading authorities like C. W. Gilmore, did not oppose such a mounting. Most significantly, over the long run (forty-five years in this case), the evidence did win out. Berman and McIntosh's study finally produced the conclusive evidence and argument that, for the paleontological community, placed the issue beyond controversy.

In any field of science, as soon as one controversy is settled, another heats up. While Berman and McIntosh were settling the long-standing headless dinosaur issue, paleontologists were engaged in a spirited controversy over dinosaur physiology. This will be the topic of the next chapter.

# 2

# THE HERESIES OF DR. BAKKER

The Age of Reptiles ended because it had gone on long enough and it was all a mistake in the first place. A better day was already dawning at the close of the Mesozoic Era. There were some little warm-blooded animals around which had been stealing and eating the eggs of the Dinosaurs, and they were gradually learning to steal other things too. Civilization was just around the corner.

—Will Cuppy, "The Dinosaur,"
in *How to Become Extinct*


The picture's pretty bleak, gentlemen.... The world's climates are changing, the mammals are taking over, and we all have a brain about the size of a walnut.

—*Stegosaur* lecturer to dinosaur audience
in *The Far Side* by Gary Larson


THE QUOTES from Will Cuppy and Gary Larson capture one popular image of the dinosaur: an evolutionary cul-de-sac, stolidly resigned to replacement by superior mammals. Combating this image of dinosaurs has been the life's work of paleontologist Robert T. Bakker, a self-described "dinosaur heretic." In 1968, while still an undergraduate student at Yale University, Bakker published an article calling for a new approach to dinosaur studies (Bakker 1968). In his subsequent articles, book, novel (*Raptor Red*), and numerous television appearances (where his long hair, beard, and wide-brimmed hat have made him almost as familiar as the dinosaurs themselves), Bakker has been the tireless champion of dinosaur revisionism.

This chapter will present Bakker's "heresies" and the debates they

## The Heresies of Dr. Bakker

provoked. As with the controversies examined in the last chapter, the debates precipitated by Bakker have drawn little attention from professional historians of science. The most extensive accounts of these debates are in such semipopular works as those by Lessem (1992), Wilford (1985), and Desmond (1975). I shall reconstruct the debates from the original articles.

According to Bakker, by 1968 dinosaurs had become synonymous with obsolescence: dimwitted hulks that managed to dominate by size alone until they were superseded by new, smarter, faster, warm-blooded mammals (Bakker 1968). He points out, though, that mammals evolved in the late Triassic at roughly the same time as the dinosaurs. Dinosaurs quickly replaced the successful mammal-like reptiles that had dominated the Triassic. Dinosaurs remained the largest and most successful land animals for the next 150 million years. Mammals remained the primitive and diminutive occupants of a few narrow ecological niches until the dinosaurs' extinction. How, Bakker asks, could the hapless, lumpish, swamp-bound leviathan of the popular stereotype have been so successful as the dinosaurs were? Bakker sought to replace the image of dinosaurs as slow, stupid, and mucking about in swamps with a view of them as fast, smart, and firmly on dry land.

Bakker never adequately documents the claim that his construal of dinosaur "orthodoxy"—slow, stupid, and swamp-bound—was the received view among paleontologists. In an essay entitled "Return of the Dancing Dinosaurs" Bakker claims that he is restoring the lively, energetic view of dinosaurs held by early paleontologists such as E. D. Cope, T. H. Huxley, and O. C. Marsh (Bakker 1987). Though his claims about Cope, Huxley, and Marsh are well documented, the alleged eclipse of this view by a stultifying "orthodoxy" is not. Some have even asserted that Bakker's broadsides against paleontological "orthodoxy" are aimed at straw men (Crompton and Gatesy 1989, p. 111).

Bakker seems to have been chiefly offended by the standard museum depictions of dinosaurs—restored as if shuffling about in an awkward semisprawl (Bakker 1987, pp. 43–44). He opens his book *The Dinosaur Heresies* with an autobiographical anecdote about a moment of deep insight that came late at night in the Yale Peabody Museum: "I remember the first time the thought struck me! 'There's something very wrong with our dinosaurs.' I was standing in the great Hall of Yale's Peabody Museum, at the foot of the *Brontosaurus* skele-

The sluggish dinosaur of "dinosaur orthodoxy." Drawing by Jessica Gwynne, from Desmond (1975).

ton. It was 3:00 A.M., the hall was dark, no one else was in the building. 'There's something very wrong with our dinosaurs.' The entire Great Hall seemed to say that" (Bakker 1986, p. 15).

Maybe Bakker's reaction against dinosaur "orthodoxy" was more a response to museum restorations and to popular stereotypes (as in the Cuppy and *Far Side* quotations) than to the paleontological literature. Of course, museum displays and even popular metaphor often reflect accepted theory, but without more citations of the paleontological literature, it seems that Bakker was too hasty in saying that his colleagues view dinosaurs as sluggish, stupid, evolutionary failures. Any claim that Bakker has brought about a "revolution" in dinosaur science (as, e.g., in Desmond 1975) must therefore be carefully qualified.

Bakker has made some genuinely controversial proposals. His most distinctive claim was made in a 1971 article in *Evolution* entitled "Dinosaur Physiology and the Origin of Mammals" (Bakker 1971). Here he introduced the view that dinosaurs were warm-blooded—as fully endothermic as any present-day bird or mammal. This is Bakker's most famous "heresy."

Some have commented on the tone of the debate over dinosaur endothermy. Wilford titles his chapter on the controversy "Hot Times over Warm Blood" (Wilford 1985). Paleontologist James O. Farlow, reflecting on the controversy in 1990, makes a slap at Bakker:

## The Heresies of Dr. Bakker

> Unfortunately, the strongest impression gained from reading the literature of the dinosaur physiology controversy is that some of the participants have behaved more like politicians or attorneys than scientists, passionately coming to dogmatic conclusions via arguments based on questionable assumptions and/or data subject to other interpretations. Many of the arguments have been published only in popular or at best semi-technical works, accompanied by rather disdainful comments about the stodgy "orthodoxy" of those holding contrary views; what began as a fresh way of considering paleontological problems has degenerated into an exercise in name-calling. (Farlow 1990, p. 44)

Bakker clearly enjoys and promotes his role as a heretic, and he does contrast his views with an alleged dinosaur "orthodoxy." Occasionally his comments *are* disdainful, as when he identifies Charles W. Gilmore as the paleontologist largely responsible for the emergence of such "orthodoxy":

> Between 1909 and 1945 Charles W. Gilmore published extensive descriptive memoirs on dinosaurs, and these studies shaped the new, narrow interpretations of dinosaurean biology. Gilmore's work had a plodding adequacy devoid of the depth of biological insight shown by Mantell, von Meyer, Huxley, and Riggs. Gilmore described bones as nearly totally inanimate creations and never displayed any first hand experience with the muscular anatomy or joint structure of extant species. Gilmore's lead-footed reconstructions became the universal standards for textbooks and museums. (Bakker 1987, p. 44)

Bakker is also intransigent; I can find no instance in which he has modified an opinion in response to criticism. He evinces complete self-assurance, and his conclusions ring with conviction. In the final chapter I shall speculate on the reasons for Bakker's tone and style. For the present, I shall focus on the arguments of Bakker and his antagonists and ignore the rhetorical and other "nonscientific" aspects of the controversy.

I do not deny the significance of rhetoric, ad hominem attacks, accusations of unprofessional behavior, and so on in determining the outcome of the debate. On the contrary, I think the testimony of Farlow and others clearly shows that such factors can have a very significant effect. Rather, I focus on the arguments because it is precisely the alleged social construction of such "rational" aspects of scientific debate that is the concern of the thinkers examined later in this book. A salient theme of the pertinent sociological analyses is that the form and content of scientific arguments are social constructs. It is not a matter of objectivity, rationality, and evidence *versus* social construction; rather, the claim is that "objectivity," "rationality," "evidence," etc., are *themselves* socially constructed through and through. (Steven Shapin makes this point very forcefully in Shapin 1992; I evaluate such claims in chapter 6.)

In the early and mid-1970s Bakker wrote the key articles advancing the case for dinosaur endothermy. By 1975 Bakker could claim in *Scientific American* that he had initiated a "renaissance" of dinosaur studies (Bakker 1975a). In that same year historian Adrian Desmond published *The Hot-Blooded Dinosaurs,* a strong endorsement of Bakker's views (Desmond 1975). The subtitle of Desmond's book was "A Revolution in Palaeontology," but if such a revolution had occurred, it was still news to most of Bakker's colleagues. Professional criticism mounted, leading the American Association for the Advancement of Science to hold a symposium in 1978 that brought together Bakker and his chief critics. The papers presented at this symposium were published in 1980 under the title *A Cold Look at the Warm-Blooded Dinosaurs* (Thomas and Olson 1980).

*A Cold Look* remains the central document of the whole controversy. Unsurprisingly, the AAAS symposium settled nothing. Bakker remains unswayed, as do his critics. According to one commentator, the debate over dinosaur endothermy reached a "dead end" with the symposium (Lessem 1992, p. 43). This is a fair assessment; the debate climaxed at the AAAS symposium, and the battle lines have changed little since. The 1978 symposium therefore provides a convenient stopping place for this case study, though at the end of the chapter I shall briefly mention interesting recent developments.

By the time Bakker published the *Scientific American* article in 1975, he had stated the essentials of his case for dinosaur endothermy. I must beg the reader's indulgence as I present rather long and unvarnished summaries of Bakker's arguments and his critics' replies during the years 1971–75.

First, some terminological confusions must be cleared up. Some of Bakker's early critics chided him for not distinguishing between homeothermy and endothermy (Feduccia 1973, pp. 166–67). "Homeothermy" refers to an animal's ability to maintain its temperatures within a narrow range. "Endothermy" refers to the thermoregulatory pattern whereby an animal generates body heat by maintaining a high level of basal metabolism (see Ostrom 1980 for clear definitions of the relevant terms). Homeothermy and endothermy are not synonymous; an active "ectothermic" (i.e., "cold-blooded") animal in a stable temperature environment could be a homeotherm.

The problem is that Bakker, at least in his 1971 article, talks much more about homeothermy than about endothermy. In response to his critics Bakker says, "I specifically stated that dinosaurs achieved homeothermy through bird-like endothermy and high yearly energy budgets" (Bakker 1974, p. 497). However, the original statements were not completely clear. In the following discussion let us assume that when Bakker and his critics use the term "homeothermy," they mean "homeothermy achieved through bird-like endothermy."

In Bakker's important 1971 *Evolution* article, the main topic is the correlation of posture to physiology. Present-day lizards and amphibians have what Bakker calls the Sprawling Gait (I follow Bakker's practice of putting the first letters of these terms in uppercase): The femur and humerus project almost horizontally, and the body can be lifted from the ground only with considerable effort from the ventral shoulder and hip muscles (Bakker 1971, p. 637). Mammals and birds, on the other hand, stand fully erect with the proximal limb segments operating almost vertically, so that body weight is supported less by muscle and more by the direct transmission of forces through bone and joint. The Sprawling Gait is fine for short bursts of activity, but not for long periods of hunting, grazing, fleeing, etc. Bakker addresses the relative success of early mammals and dinosaurs by examining the joint and postural mechanics of numerous living and fossil tetrapods.

Dinosaurs and mammals must have had a common reptilian ancestor at some point in the early Carboniferous (p. 638). The starting point for tracing the physiological evolution of dinosaurs and mammals is the reconstruction of the locomotion and thermoregulation of the earliest reptiles. Bakker concludes that the very primitive Carboniferous reptiles were typical sprawlers that were short-limbed and probably slow moving (pp. 639–40).

Among the genera of living lower tetrapods, those with the shortest limbs are consistently those with the lowest activity temperatures, that is, those that pursue their ordinary activities at the lowest body core temperatures (pp. 640–42). Present-day short-legged salamanders and reptiles are mostly nocturnal or crepuscular and seek shaded, moist areas during the day. A low activity temperature is a useful adaptation for such a lifestyle. Bakker concludes that the earliest Carboniferous reptiles, the ancestors of dinosaurs, lizards, and mammals, had low activity temperatures (p. 643).

Triassic reptiles called "thecodonts" are believed to have been the immediate ancestors of the dinosaurs. By the Mid-Triassic Period, some thecodonts had adopted what Bakker calls the Semi-Erect Gait. The body proportions of the semi-erect thecodonts resembled those of present-day monitors (like the Komodo dragon). Monitors are very active hunters with a high (37° C) activity temperature and with considerable capacity for generating endogenous heat. The semi-erect thecodonts, says Bakker, were built for even more active hunting than were monitors. Monitors are still sprawlers; these hunting thecodonts stood much more erect and possessed lower limb bones as long or longer than the upper limb bones—a cursorial feature not found in monitors. Bakker concludes that these thecodonts had an activity level and capacity for generating endogenous heat as great or greater than that of monitors. With dinosaurs we finally get the Fully Erect Gait. The postural mechanics of dinosaur fore and hind limbs is virtually identical to that of marsupial and placental mammals (p. 645).

According to Bakker, the joint angulation and limb proportions of dinosaurs show that, like the present-day higher mammals, they were diversified into a variety of locomotor types (p. 649). Some, like *Stegosaurus,* were amblers like elephants; others, like *Triceratops,* were gallopers like rhinos. Bakker claims that dinosaurs were actually as well or even better adapted for continuous activity and rapid movement as the largest land mammals (pp. 647–49). He concludes:

> A radiation based on fully erect locomotor types including both large ambling and large, fast-running types probably reflects the high activity levels of homeothermy. Fully erect limbs with relatively straight joints permit the movement of heavy weights over very long distances with a minimum of effort. . . . In both elephantine and rhino locomotor types the basic Fully Erect Gait permits rather continuous activity. All the sprawling tetrapods, living and fossil, put together do not show the great variety of joint angulation and limb proportions seen either in therians or in dinosaurs. The activity patterns reflected in dinosaur locomotion clearly are much closer to those of therians than those of lizards. (p. 649)

## The Heresies of Dr. Bakker

Bakker believes that he has traced the evolution of dinosaurs from nocturnal crawlers to diurnal gallopers. He infers that the evolution of posture and probable activity levels in the dinosaur lineage from the early Carboniferous to the Mid-Mesozoic reflects profound physiological changes. This seems to be a reasonable conclusion and is supported by further anatomical and evolutionary considerations.

For instance, dinosaurs possessed a number of adaptations for a high-energy lifestyle. They possessed formidable dental and digestive apparatus (Norman 1991, pp. 109–13). Their large size would provide insulation against heat loss (even dinosaur hatchlings were rather large). In fact, a large, active animal in a warm climate often faces the problem of discharging excess heat. Birds, for instance, expel excess heat through an air-sac system that is connected to the lungs and extends deep within the body.

Bakker interprets large cavities and pores in the vertebrae of dinosaurs as evidence that dinosaurs possessed an air-sac system (Bakker 1971, p. 650; 1972, p. 81). Some of Bakker's critics, such as Alan Feduccia, charge that it is mere conjecture to regard these cavities as part of an air-sac system (Feduccia 1973, pp. 167–68). The standard view is that these pleurocoels and pneumatopores are adaptations that serve to lighten the enormous weight of the skeleton (McIntosh 1990, p. 359). Bakker's reply to Feduccia's criticism on this point is little more than a dogmatic reassertion of the original claim (Bakker 1974, p. 499; see Reid 1997, pp. 466–67, for an argument supporting Bakker's air-sac hypothesis and Ruben et al. 1997 for a contrasting view of dinosaur respiration).

Finally, Bakker claims that the immediate ancestors of dinosaurs, and the dinosaurs themselves, were extremely successful in displacing or severely limiting their competitors. The sprawling mammal-like therapsids of the Mid-Triassic were replaced by semi-erect and fully erect archosaurs by the end of the period (Bakker 1971, p. 654). Fully erect mammals did not—Bakker thinks *could* not—radiate into large body-size niches until the dinosaurs had become extinct (p. 656). He repeats the claim in later writings that fully erect and endothermic mammals could have been kept in small-body and probably nocturnal niches only by endothermic dinosaurs (Bakker 1975a, 1986).

How good is Bakker's case so far? Peter Dodson has recently challenged the claim that large ceratopsids such as *Triceratops* were well

The lean, mean dinosaur of Bakker's vision. Drawing © 1985 by Robert F. Walters.

designed for running, much less galloping. He argues that anatomical and trackway evidence requires a sprawling stance for the front limbs (Dodson 1996, pp. 273–76). He contrasts the anatomy of large ceratopsids with modern cursorial mammals:

> In running mammals, especially perissodactyls (horses) and artiodactyls (deer and horned stock such as antelopes and gazelles), the feet are elongated, the proximal limb segments (humerus, femur) are short, the chest is narrow, the shoulder moves freely, and the upper end of the humerus is narrow. Running mammals have eliminated the clavicle a bone that . . . supports and separates the shoulder girdles and limits their mobility. . . . In none of these respects do ceratopsids emulate running mammals. In ceratopsids, the feet are short, the humerus and femur are strikingly long compared to the ulna and tibia, and the upper end of the humerus has a very broad crest for the attachment of breast muscles. A broad pectoral crest is generally found in animals that sprawl but not in those that run. Mobility of the shoulder is especially suspect. (pp. 277–78)

Bakker vigorously rebutted earlier such criticisms (Bakker 1974), but it is clear that his assertions about the cursorial capabilities of ceratopsids are still much disputed by leading experts (Forster and Sereno 1997, p. 327, concur that ceratopsids were "not built for speed").

The main facts of Bakker's natural history are not in dispute: Therapsids really did decline as the archosaurs rose into prominence in the Triassic. Mammals remained ecological bit players until the dinosaurs' demise. Whether these facts of natural history admit of interpretations other than Bakker's is, of course, another question. For instance, M. J. Benton (1997, pp. 211–12) argues that dinosaurs did not drive out the mammal-like reptiles, rhynchosaurs, and thecodontians. Rather, dinosaurs simply occupied the niches left vacant when those earlier creatures were extinguished in a mass-extinction event. Paul Sereno, in a review article on dinosaur evolution in *Science,* notes that there was a fifteen million year gap between the initial radiation of dinosaurs and their rise to global dominance. He concludes that opportunistic replacement is a more likely hypothesis than is competitive displacement (Sereno 1999, p. 2137).

The critical response to Bakker's article came quickly. A reply to Bakker by Alan Feduccia and another coauthored by Albert F. Bennett and Bonnie Dalzell soon appeared in *Evolution.* Bennett and Dalzell argue as follows:

> Bakker fails to demonstrate a logical connection between fully erect posture and homeothermy. The essence of his argument is that since the only vertebrates now possessing a fully erect gait are homeothermic and since dinosaurs are hypothesized to have possessed an upright posture, dinosaurs must have been homeothermic. Such a statement is neither logically nor biologically sound. If there is a causal connection (implication) between these factors, what is its physiological basis? If no causal linkage can be demonstrated, then the two components of the conjunction must approximate a full correlation for the argument to be compelling. (Bennett and Dalzell 1973, p. 170)

Feduccia similarly criticizes the thesis of dinosaur endothermy, a thesis he attributes to John Ostrom as well as to Bakker:

> The major thesis of the Ostrom-Bakker argument relies upon the fact that there are no extant terrestrial vertebrates with erect posture that are ectothermic. The conclusion is that all erect terrestrial vertebrates must therefore be endothermic. It is true that there are no erect ectotherms in the modern fauna, but the converse does not hold; that is, animals which do not have the erect posture should not be capable of muscular thermogenesis sufficient to maintain an endothermic regulatory system. Thus it is surprising that cetaceans, pinnipeds, dugongs, and other aquatic, or partially belly-resting modern mammals (echidnas, moles, etc.) have physiological temperature regulation. (Feduccia 1973, p. 167)

These criticisms are somewhat confused. Bennett and Dalzell seem to conflate the logical relation of implication with the physical one of causation. To say that p implies q is merely to say that, as a matter of fact, we never have p without q. This, of course, is not the same thing as saying that p causes q. Also confused is the charge that Bakker fails to demonstrate a "full correlation" between endothermy and the possession of an erect posture. Presumably, this means that if Bakker cannot establish a causal connection between endothermy and erect posture, he must show that among extant animals all and only endotherms possess such a posture. Feduccia implies this same demand when he concedes that all extant erect animals are endotherms but then lists various sorts of nonerect endotherms (pinnipeds, cetaceans, etc.) to show that Bakker cannot claim the converse.

Even if it were the case that among living animals all and only endotherms possessed an erect posture, it is hard to see how this would provide a compelling argument for dinosaur endothermy. If, contrary to fact, it had turned out that erect posture and endothermy were perfectly correlated among living fauna, this might merely be an accidental result of evolutionary history—erect ectotherms and nonerect endotherms simply having failed to survive. By comparison, having feathers and having a furcula (wishbone) are perfectly correlated in living fauna, but feathers could well have evolved before wishbones (see below).

Bakker's critics also seem uncertain as to whether he is arguing that endothermy is necessary, sufficient, or both necessary and sufficient for an erect posture. Such confusion is understandable; Bakker's argument connecting endothermy and erect posture is expressed somewhat circuitously, and its relevance to his case for dinosaur endothermy is not clear.

Bakker concedes that small (under 10 kg) terrestrial endotherms might do well enough with a sprawling posture (Bakker 1971, p. 637). However, the weight-bearing capacity of bone increases with respect to the square of linear dimensions, and body weight increases with respect to their cube, so sprawling larger animals will experience greater strain than small ones. Bakker concludes that a sprawling posture might allow a large terrestrial vertebrate to engage in short bursts of movement but not long periods of continuous activity. Endotherms have an extremely high energy budget and so, Bakker assumes, must stay continuously active to forage or to hunt.

## The Heresies of Dr. Bakker

To say that endotherms must stay continuously active is clearly an exaggeration. Surely Bakker has noticed how much time lions, for example, spend sleeping and lounging. In fact, the lives of most large terrestrial vertebrates are like the proverbial life of the infantry soldier —long periods of boredom punctuated by moments of sheer terror. Perhaps Bakker is merely saying that endotherms are capable of much longer periods of sustained activity. For instance, the Komodo dragon (an ectotherm) is an ambush hunter incapable of running down prey over long distances like (endothermic) wolves or cape dogs. Bakker's argument may be set out semiformally as follows:

1. Endotherms have high energy budgets.
2. Animals with high energy budgets must have high activity levels.
3. Therefore, endotherms must have high activity levels.
4. If an animal has a high activity level, if it is a large terrestrial vertebrate, it must have an erect posture.
5. Therefore, if an animal is an endotherm then, if it is a large terrestrial vertebrate, it must have an erect posture.

As it stands, the whole argument is beside the point. Bakker's apparent aim is to use dinosaur posture as evidence for dinosaur endothermy, but the above argument works the other way around, inferring erect posture from the energy demands of endothermy. Given Bakker's apparent aims, it is blatantly question-begging to start by assuming endothermy. Besides, even if all large terrestrial endotherms possessed an erect stance, the converse, which is what Bakker must show, need not hold.

Perhaps we should construe the argument as offering an inference to the best explanation. Bakker has argued that from the Carboniferous to the Mesozoic the creatures in the lineage leading to dinosaurs developed an ever more erect posture. Perhaps the best explanation for the development of the Fully Erect Gait in the dinosaurs is that they are the end result of an evolutionary process whereby increasingly endothermic physiologies made increasingly upright postures selectively advantageous. So construed, the causal argument, expressed semiformally above, is a way of fleshing out this explanatory hypothesis by showing that endothermy would explain the trend toward more erect stances. More precisely, the datum to be explained is the evolutionary trend toward more erect posture in the ancestors of dinosaurs;

the hypothesized best explanation is that increasingly endothermic physiologies made such postures increasingly selectively advantageous.

So construed, Bakker's argument does not beg the question, but it faces other serious problems. For instance, Paul Sereno has argued (1991) that a semi-erect stance is not a step on the way to a fully erect posture. More fundamentally, what justifies the inference to endothermy as the best explanation of dinosaur posture? The appeal to analogies among extant animals is weak. The failure to find erect ectotherms among extant fauna is no reason think that they might not have existed previously in evolutionary history. Present-day correlations between erect posture and endothermy therefore cannot be projected into the evolutionary past. Why then should we prefer Bakker's inference as the best explanation when other hypotheses are prima facie equally or more plausible?

For instance, Feduccia suggests that upright posture may have been an adaptation necessitated simply by the dinosaurs' very large size (Feduccia 1973, p. 167). *Diplodocus* could not have been a sprawler; it is a biomechanical impossibility. Would not an upright posture be a selective advantage to *any* large terrestrial animal, whatever its thermoregulatory physiology? As Bennett and Dalzell point out, any large terrestrial animal would experience excessive bone shearing forces in a sprawling stance; an erect posture would be an advantage to any such creature (Bennett and Dalzell 1973, pp. 171–72). More recent discoveries have provided an effective reply: The earliest known dinosaurs, such as *Eoraptor, Staurikosaurus,* and *Herrerasaurus,* were slender, lightweight creatures, yet they already evince the upright, typically dinosaurian posture (Benton 1997, p. 208). So the posture seems to have come first, the enormous size later.

At the time, though, Bakker's critics may have felt that he had given them a mere scenario: Endothermy appeared to be only one of several plausible explanations of dinosaur posture. Notoriously, such evolutionary scenarios easily degenerate into "just so" stories. Just so stories provide an imaginative and appealing account of how something *could* have happened (e.g., dinosaur endothermy making erect posture selectively advantageous), but they fail to provide evidence for favoring one such story over others. Bakker's critics, whatever their points of confusion, were justified in demanding more evidence.

Bakker seems aware that his hypothesis needs more support since most of his reply is devoted to two arguments not mentioned in his

1971 *Evolution* article. These arguments are presented in a 1972 *Nature* article and elaborated in subsequent writings (Bakker 1972, 1974, 1975a, b, 1980, 1986).

The first argument concerns differences in bone histology between endotherms and ectotherms. Mammalian and avian bone very often has densely packed vascular structures called Haversian canals. According to Bakker, reserves of bone mineral are necessary for the maintenance of internal chemical homeostasis. Endotherms, because their metabolic levels are continuously adjusted and vary enormously between rest and maximum energy output, must have rapid access to bone minerals. A high degree of bone vascularization allows rapid mineral transfer from bone to plasma (Bakker 1972, p. 82).

The bone of living and fossil reptiles, with the exception of one of the fossil mammal-like reptiles, is poorly vascularized. According to Bakker, some mammals possess a bone architecture called laminar bone that permits an even closer association of bone cells and capillaries than does Haversian bone. Bakker asserts that no living reptile possesses laminar bone, but it was present in another of the fossil mammal-like reptiles (ibid.).

Bakker reports that all of the dinosaurs which he has investigated possessed either Haversian or laminar bone or both (ibid.). His conclusion is that the affinities in bone histology between dinosaurs and birds and mammals is evidence for dinosaur endothermy. The presence of highly vascularized bone in dinosaurs is certainly very interesting and suggestive. Nevertheless, as we shall see, the precise function or functions of Haversian bone, and hence its relation to thermoregulatory physiology, are not fully understood (de Ricqlés 1980, p. 122).

Perhaps Bakker's most interesting argument has to do with predator and prey ecology as it relates to endothermy and ectothermy. A very clear exposition of this argument is given in his 1975 *Scientific American* article. A key indicator of fossil metabolism is the ratio of the "standing crop"—the total biomass of predator to prey in a fossil community (Bakker 1975a, p. 61). If prey populations remain stable, they will not support a standing crop of predators larger than a certain size. Since the energy budgets of ectotherms are so much lower than those of endotherms, a given stable prey population will sustain a much higher standing crop of ectothermic than endothermic predators. For instance, thirty tons of prey carcasses a year will sustain a standing crop of forty tons of Komodo dragons, but only 2.5 tons of lions. Similar ratios hold whether size of predators and prey is large or small, says Bakker (p. 62).

Bakker next provides quite detailed information on predator-prey ratios in a number of fossil communities. The earliest land predators capable of killing large prey were Permian finback creatures such as *Dimetrodon* (p. 64). *Dimetrodon* and its kin (called "sphenacodonts") evinced a very primitive limb anatomy and a distinctively ectothermic bone histology with few Haversian canals, poor vascularization, and distinct seasonal growth rings (indicating, as is typical of ectotherms, that it grew faster in wet seasons than in dry ones). Also, sphenacodonts and their prey were confined to areas of warm and equable temperatures near the Permian equator (p. 66).

At this point a skeptic might raise the ticklish problem of how we know that the ratios present in a fossil assembly are representative of the living community (we shall see that Bakker's critics do raise this criticism). Bakker anticipates this objection by claiming that the fossil samples of sphenacodont communities are quite extensive and that the counting techniques employed give a representative ratio of predator to prey biomass for the living animals (p. 67). He concludes that predator to prey ratios for such communities are quite high— from 35 to 60 percent (p. 68).

In the late Permian the dominant land animals were the mammal-like therapsids. These creatures are called "mammal-like" because from the late Permian to the Mid-Triassic one line of therapsids increasingly resembled mammals in numerous details of skeleton and tooth anatomy. Bakker claims that advanced therapsids were actually difficult to distinguish from the first true mammals. The bone histology of the therapsids is heavily vascularized, indicating that the transition from ectothermy to endothermy came early and suddenly. A further indication of therapsid endothermy is that they colonized high latitudes at a time when there were rather severe latitudinal temperature gradients (ibid.). Predator to prey ratios were low, but still about three times higher than those found in today's advanced mammalian communities (p. 69). The picture that emerges is that therapsids were moderate endotherms with a heat production between that of lizards and modern mammals.

By the Mid-Triassic the therapsids had been largely replaced by the immediate ancestors of the dinosaurs, the thecodonts. According to Bakker, advanced thecodonts had fully "endothermic" bone

# 37
# The Heresies of Dr. Bakker

Robert Bakker, dinosaur heretic.

histology (p. 70). Information about predator-prey ratios for thecodonts is scanty, but thecodonts were widely distributed geographically, even into the cooler regions of southern Gondwanala.

Dinosaurs were widely distributed geographically, but this is not very good evidence for dinosaur endothermy since equable climate conditions prevailed worldwide during the Jurassic and Cretaceous.

The other evidence is much less equivocal, says Bakker. He claims that the bone in all dinosaur species is fully endothermic, and that some species had blood vessel concentrations higher than those of living mammals. Further, the predator-prey ratios for dinosaurs in the Triassic, Jurassic, and Cretaceous are usually from 1 to 3 percent, fully as low as the ratios for advanced mammalian fossil communities in the Cenozoic (p. 71).

Bakker provides an impressive array of arguments in which a number of independent lines of evidence are adduced in favor of dinosaur endothermy. All of this evidence and more was addressed by Bakker and his antagonists at the 1978 AAAS symposium, to which we now turn.

Two of the papers presented attack Bakker's argument based on predator-prey ratios. One is by James O. Farlow, and the other is co-authored by Pierre Béland and D. A. Russell. Farlow's paper is the clearer and more comprehensive of the two, so I shall focus on his critique.

Farlow begins with an examination of energy budgets and predator-prey biomass ratios in modern animal communities, data upon which Bakker's argument is crucially dependent (Farlow 1980, pp. 57–67). He notes that some studies do report cases in which an endothermic predator requires far more food than an ectothermic predator of the same weight (p. 57). However, it is not clear that such ratios would hold over a wide range of predator body sizes. The mere fact that so few modern animal communities have really large terrestrial ectothermic predators means that there is little information here.

Attempts to fill in the gap by comparing large mammals to small ectotherms lead to numerous difficulties in interpretation (p. 61). For instance, because of their short life spans and rapid rate of reproduction, populations of small animals have a much higher rate of biomass turnover than do populations of large animals. Higher turnover rates should support a larger biomass of predators, so extrapolation from the predator-prey biomass ratios of communities with small ectothermic predators to a hypothetical community of large ectothermic predators might overestimate such ratios for the latter. So Bakker's prediction of predator-prey ratio in communities of large ectotherms could well be too high.

A crucial assumption of Bakker's argument is that predator-prey biomass ratios are regulated solely by the amount of food produced

## The Heresies of Dr. Bakker

reliable procedures in making these calculations. Thus, Bakker devotes considerable space in his long essay in *A Cold Look* to the defense of the taphonomic procedures he employed in making the biomass calculations (Bakker 1980, pp. 437–45). Unfortunately, even the best such procedures provide no way of responding to point 3 in the critics' case.

Let us assume that the taphonomic procedures used by Bakker did not automatically slant the calculations in his favor. How could his expectations hold good over so many and diverse fossil communities unless his fundamental assumptions were essentially correct? If, for instance, the order-of-magnitude projected difference in biomass ratios is not a sound expectation, why do such ratios consistently turn up in fossil communities? To give teeth to this reply, extensive statistical testing is needed to show just how closely the data conform to Bakker's expectations and just how unlikely it is that the results are the effects of stochastic noise introduced by factors such as those adduced by Bakker's critics. For instance, such tests would have to adjust for the factors mentioned in the critics' point 3. Bakker's critics appear justified in demanding further testing.

No one argument or set of arguments was decisive for the outcome of the 1978 AAAS symposium. At no point was Bakker simply checkmated; he was left with plenty of room for maneuver. Nothing rationally compelled the skeptical outcome. So was the continued skepticism of the vertebrate paleontologists simply a reflection of prejudice?

Skepticism toward new ideas almost always involves a considerable amount of sheer epistemic inertia; sometimes a great impulse is needed to move the bulk of received opinion. Still, in epistemic matters, the fundamental rule of conservatism applies: "If it ain't broke, don't fix it." What appears to the radical as sheer prejudice, knee-jerk hostility to innovation, is often really a habitual response conditioned by the hard school of experience. Most new ideas are bad, so the few good ones really have to prove themselves. Of course, epistemic conservatism, like political conservatism, can become extremist and irrational.

Prima facie, therefore, the continued skepticism of the scientific community toward Bakker's hypothesis was a collective decision based on the *rational* weighing of all the lines of argument and evidence. Bakker's evidence from bone histology was intriguing, but, for the reasons given by de Ricqlés, not decisive. Likewise, the predator-prey biomass arguments were ingenious, but, for the reasons given by Farlow and others, far from conclusive. Epistemological conservatism places the burden of proof on the innovator, and this is right and proper. In the minds of his colleagues, Bakker's arguments and evidence would not bear that burden.

Though the intense debate over dinosaur endothermy waned soon after the publication of *A Cold Look,* discussions of dinosaur physiology continue (see the relevant articles in Farlow and Brett-Surman 1997 and chapter 14 of Fastovsky and Weishampel 1996 for detailed surveys; Varricchio 1998 provides a quick overview). For instance, R. E. H. Reid and A. Chinsamy have demonstrated, *pace* the earlier claims of de Ricqlés, that lines of arrested growth (LAGs) do occur in dinosaur bone (Chinsamy 1993; Reid 1987, 1990). LAGs in a bone indicate that bone growth stopped for a period. In extant ectotherms the pattern of LAGs can indicate fluctuations in growth due to seasonal temperature variations; growth stops in cool seasons and continues in warm. Endotherms, with their much greater independence from climatic influence, continue growing right through the winter and show no well-developed LAGs.

Reid and Chinsamy regard the LAGs in dinosaur bone as representing yearly increments of growth, showing that—like reptiles and unlike mammals and birds—dinosaurs were susceptible to seasonal changes. Dinosaur bone is therefore a paradox; dense vascularization indicates rapid growth and high activity levels, but annual LAGs indicate a continued sensitivity to seasonal temperature changes. The bone evidence therefore indicates that dinosaur metabolism may have been intermediate between the "good reptile" and the mammalian or avian extremes.

Another interesting development is the recent discovery of feathered dinosaurs and reptiles. In a 1998 *Nature* article, Ji Qiang, Philip J. Currie, Mark A. Norrell, and Ji Shu-An reported the discovery in northeastern China of two feathered Upper Jurassic/Lower Cretaceous theropods (Ji et al. 1998). This claim seemed to confirm the long debated (and still hotly disputed) theory of bird descent from dinosaurs. If dinosaurs and birds *are* closely related, it may be easier to argue that they shared similar metabolisms, but there are many uncertainties and potential pitfalls in such arguments.

A line of evidence that seems to give a clearer answer comes from John Ruben and Willem Hillenius and their colleagues (Ruben et al. 1997). They note that endothermic animals have a high rate of respiration and that this poses a danger of excessive water loss since inhaled air would evaporate water from the lungs. To compensate, endotherms (and only endotherms) have developed nasal turbinates, bones in the nasal cavity which provide a large surface area for moisture-reclaiming mucous membranes. Unlike other supposed indicators of "warm-bloodedness" (e.g., erect posture, densely vascularized bone), nasal turbinates seem to have an unambiguous and exclusive functional significance for endothermic organisms: That is, all endotherms —and only endotherms—have a need for respiratory turbinates, so they should be present in endotherms and lacking in nonendotherms. Since dinosaurs have no respiratory turbinates, these authors conclude that dinosaurs were ectothermic, or nearly so, during their periods of routine activity (p. 516).

Unfortunately for those who yearn for closure in the dinosaur physiology debate, every piece of apparently conclusive evidence must be balanced with opposing indicators. Reese E. Barrick, Michael Stoskopf, and William J. Showers (1997) argue that the ratio of oxygen isotopes in dinosaur bone is evidence of endothermy. Bone contains phosphate ($PO_4$), and the oxygen in the phosphate comes in the form of two different isotopes, $^{18}O$ and $^{16}O$. The ratio $^{18}O:^{16}O$ in bone phosphate is partially determined by the body temperature of the animal at the time the phosphate was formed. When sufficient amounts of the original bone phosphate remains in fossils, the body temperatures of extinct animals can be estimated from that ratio.

Barrick et al. applied their isotope analysis to the fossils of a Cretaceous varanid lizard (presumably a "good reptile" ectotherm) and found that its body temperature fluctuated seasonally by at least 10° to 15° C (Barrick et al. 1997, p. 484). Dinosaurs, on the other hand, evinced little seasonal temperature fluctuation (no more than 2° C), indicating that they were homeotherms, that is, that their body temperature was maintained within a fairly narrow range in all seasons of the year.

Recall that earlier I pointed out that homeothermy is not the same as endothermy. Some researchers have argued that dinosaurs could maintain their body temperatures within a narrow range (homeothermy), not by maintaining a high metabolic level (endothermy) but simply by being big (gigantothermy). The larger the animal, the more volume it has in relation to its surface area. Since body heat is radiated from the surface, creatures with a low ratio of body volume to surface area, like mice, have to maintain a very high metabolic rate to keep their core temperatures high enough. Elephants have the opposite problem; their massive bulk produces so much heat that it must be dissipated. Large ears increase the surface area for heat radiation. Some researchers have suggested that dinosaurs, often as large and sometimes much larger than African elephants, could not have been endotherms since their food needs and heat stress would have been excessive (Paladino, Spotila, and Dodson 1997, p. 500). Barrick and his coauthors reply that even juvenile dinosaurs, who were not so large, show little seasonal isotopic variability and so must have maintained homeothermy with relatively high metabolic rates (Barrick, Stoskopf, and Showers 1997, p. 486).

So, though we have found much new evidence since the late 1970s, there are yet no conclusive answers about dinosaur physiology. (As of this writing, the *very* latest claim is the purported discovery of a fossilized, four-chambered, that is, mammal-like, heart in a *Thescelosaurus*.) Since this chapter opened with a quotation from Gary Larson's *The Far Side,* it is appropriate to conclude with another. In this cartoon an intrepid time-traveling scientist, brandishing a huge rectal thermometer, approaches the backside of a large sauropod. The caption reads: "An instant later, both Professor Waxman and his time machine are obliterated, leaving the coldblooded/warmblooded dinosaur debate still unresolved." Alas! We have no time machines, so the complicated question of dinosaur physiology may elude resolution indefinitely.

One final observation: Despite the rejection of Bakker's claim that all dinosaurs were true endotherms of the mammalian or avian sort, paleontological consensus did not revert to the "good reptile" model of ectothermic dinosaurs. R. E. H. Reid has recently defended the view that dinosaurs possessed "intermediate" thermoregulatory strategies not found in contemporary organisms. After a thorough review of bioenergetic principles, and the anatomical, histological, and other evidence, Reid concludes:

> It seems fairly likely that the great success of dinosaurs in their time need not mean that they were endotherms. They could instead have been sub-endothermic "super-reptiles," or "super-gigantotherms," with a more elevated circulatory system than any modern reptile, and no true modern physiological counterparts. Together with inherited upright limbs, which would have allowed them high mobility and pre-adapted them to support massive weights, this could

## The Heresies of Dr. Bakker

have made them superior as large terrestrial animals to all possible Mesozoic competitors. (Reid 1997, p. 471)

So Bakker may have won a sort of backhanded victory. Though his endothermy hypothesis did not prevail, the stereotype of the dinosaur as swamp-bound sluggard has not survived. No more will dinosaurs be seen as predestined failures, but as the marvelously adapted and successful creatures they were. Their very success heightens the mystery of how dinosaurs suffered such a complete and apparently sudden extinction. This issue takes us to the topic of the next chapter.

# 3

# THE "CONVERSION" OF DAVID RAUP

The second angel blew his trumpet, and something like a great mountain, burning with fire, was thrown into the sea; and . . . a third of the living creatures in the sea died.
—*The Revelation to John* 8: 8–9

Sometimes you have a really bad day, and something falls out of the sky.
—Walter Alvarez

WHAT HAPPENED TO the dinosaurs? Despite the fondest wish of many an eight-year-old, they are all gone. Well, maybe not. In their book *The Mistaken Extinction,* Lowell Dingus and Timothy Rowe (1997) argue that they continue to flourish—as birds. Still, for the nonavian dinosaurs, the Late Cretaceous was the final act, the end of the 160 million year Age of Dinosaurs. The issue of dinosaur extinction has been the topic of much speculation, some of it quite silly, and much, sometimes nasty, debate.

For approximately fifteen years, from 1980 to 1995, scientists from various disciplines engaged in a vigorous, sometimes rancorous, debate over dinosaur extinction, a debate even more intense than the warm-blooded dinosaur controversy of the 1970s. Since the mid-1990s the dispute has become less polarized as participants find more common ground. The debates over dinosaur endothermy were chiefly a preoccupation of vertebrate paleontologists. The extinction imbroglio involved physicists, astronomers, geochemists, geologists, paleontologists, and a variety of representatives of other disciplines and subdisciplines.

# The "Conversion" of David Raup

According to William Glen's count, by 1994 the number of publications contributing to this controversy had exceeded twenty-five hundred (Glen 1994, p. 2).

The recent debate began on June 6, 1980, when a paper authored by Nobel Prize–winning physicist Luis Alvarez, his son geologist Walter Alvarez, and the nuclear chemists Frank Asaro and Helen Michel appeared in the journal *Science* (Alvarez et al. 1980). This paper, "Extraterrestrial Cause for the Cretaceous-Tertiary Extinction," marks a true watershed in the history of the earth sciences. All discussion of dinosaur extinction, indeed, all discussion of mass extinction in general, can be divided into the periods before and after the publication of this paper. Prior to 1980, theories of dinosaur extinction ranged from the plausible (but weakly supported) to the speculative to the bizarre (see Jepson 1964 for an amusing account of the lack of progress made by dinosaur extinction theories up to that time). The paper by Alvarez et al. presented a clear hypothesis that made bold, testable predictions.

The startling hypothesis of the Alvarez group was that the Cretaceous-Tertiary (K/T) mass extinctions were caused by the impact into the earth's surface of an asteroid or comet of approximately ten kilometers in diameter (Alvarez et al. 1980).[1] Such an impact would release the explosive energy of an estimated one hundred million megatons of TNT—several orders of magnitude greater than the total megatonnage of all presently existing thermonuclear weapons (Alvarez 1987, p. 259). The projected effects of this impact were deemed severe enough to cause a worldwide environmental crisis, perhaps even the cessation of photosynthesis due to the darkness caused by a globe-enshrouding cloud of dust and ash (Alvarez et al. 1980). The image of a dinosaur doomsday, the destruction of the lords of the Mesozoic in a stupendous blast and the freezing darkness that followed, clearly made the impact hypothesis uniquely exciting.

Naturally, the Alvarez paper drew fire and was followed by a spate of rebuttals and replies. Glen does an excellent job of tracing the intricate twistings and turnings of the debate from 1980 to 1994 (Glen 1994, pp. 7–38; see also his 1996 update). It is not necessary to follow him through these convolutions, only to note that, by the mid-1990s, two opposing catastrophist theories had become fairly deeply entrenched, each postulating a different model of sudden K/T extinction: the bolide theory, by then often modified into a theory of mul-

Luis (*left*) and Walter Alvarez (*right*) at the K/T boundary layer. Courtesy, the Lawrence Berkeley National Laboratory.

## The "Conversion" of David Raup

tiple smaller impacts, and the volcanic theory. The volcanic theory holds that the K/T mass extinctions were caused by cataclysmic volcanic explosions.[2]

Old-fashioned, gradualist extinction theories, such as those postulating gradual cooling or receding sea levels, had to accommodate data such as the anomalously high abundance of siderophile elements and the presence of "shocked" quartz at the K/T stratigraphic horizon. Siderophile ("iron-loving") elements, such as iridium, are presumed to be abundant in the earth's iron core, but are very rare in the crust. Some meteorites have relatively high abundances of these elements, as does the magma released from some volcanoes. "Shocked" minerals are those that display characteristic deformations that are caused only by very high impacting pressures—like those produced by bolide impacts, thermonuclear explosions, and volcanic explosions. Some gradualists therefore came to admit that a bolide strike or strikes had occurred at the end of the Cretaceous, but they held that the dinosaurs were already extinct (or largely so) by then (see Archibald 1996 and 1997 for a theory that is basically gradualist but with some concessions to catastrophism).

David Raup, one of the world's leading invertebrate paleontologists, has been a major player in the mass extinction debates. Here I tell the story of Raup's "conversion" to mass extinction hypotheses. Raup gives his own version of this story in *The Nemesis Affair*, an excellent popular account of the early extinction debates and his own part in them (Raup 1986).

Raup depicts himself as an early critic of the impact scenario and as strongly inclined to dismiss its adherents without a fair hearing. As a referee for the journal *Science,* he recommended against the publication of the 1980 paper by Alvarez et al. that precipitated the mass extinction controversies with its bolide impact hypothesis. Just a couple of years later, though, Raup had become one of the chief proponents of impact theories—in fact, an advocate of a radical version of such a theory.

This story sounds like the sudden, dramatic, and comprehensive changes in theoretical commitment discussed by Thomas Kuhn in his classic study *The Structure of Scientific Revolutions* (Kuhn 1970). Kuhn obscurely but provocatively characterized such changes in terms of "conversions" and "gestalt switches." Such terminology highlighted his view (as Kuhn is usually interpreted) that scientists who undergo a radical shift in their theoretical commitments experience something like a religious conversion—they adopt a whole new set of values, aims, and practices. Like the religious convert, the scientific convert sees things in a whole new way; his or her whole worldview is changed.[3]

The merits and demerits of Kuhn's views have been exhaustively debated elsewhere (see Lakatos et al. 1970; Toulmin 1972; Laudan 1977; Newton-Smith 1981; Scheffler 1982; Siegel 1987; Hoyningen-Huene 1993; Horwich 1993; and others far too numerous to mention). Despite these vigorous, sometimes polemical, commentaries and criticisms, Kuhn-style analyses have flourished in much recent history and sociology of science. One such work is William Glen's "How Science Works in the Mass-Extinction Debates," which is the second chapter in the volume he edited in 1994, *The Mass Extinction Debates: How Science Works in a Crisis* (Glen 1994).

Here I summarize Glen's views on "conversions" in the mass extinction debates and then examine Raup's own account to see how well it accords with the Glen-Kuhn analysis. I find that Raup's published work on mass extinctions does not reflect a wholesale shift in the methodologies he adopts, the standards of theory choice he employs, or any other sort of radical reorientation toward his discipline. In other words, Raup accepts the bold new impact theory, and becomes the chief proponent of a radical version of such a theory, without significantly altering the basic aims, methods, or standards of his science. The upshot is that a close examination of Raup's "conversion" does not support Glen's Kuhnian account of radical theory change, but rather supports Stephen Toulmin's view in his 1972 book *Human Understanding* that stressed the continuity of methodology within a disciplinary matrix.

Glen's stated aim is to show how alleged "subjectivities," such as factors of personal biography, the prestige of disciplinary doyens, and philosophical tenets, interact with "objectivities," such as, presumably, hard data (Glen is not very clear on what he counts as an "objectivity"), to condition the workings of science in the mass extinction debates (Glen 1994, p. 3). Part of his analysis focuses on scientific "conversions."

According to Glen, most of the scientists involved in the mass extinction debates quickly declared for or against the impact hypothesis, dividing up strictly along disciplinary lines, and have not altered their

## The "Conversion" of David Raup

attitudes (pp. 50–55, 76–77). Glen analyzes the rare exceptions to this rule in a section entitled "Gestalts, Polemics, and Conversions" (pp. 72–79). Here he notes that the impact hypothesis confronted earth scientists with some startling new claims (pp. 78–79).

Glen indicates that the suddenness of the extinctions proposed by the impact theory went strongly against the grain of the working earth scientist of 1980. Martin Rudwick, in an interview with Glen, remarked that as an invertebrate paleontologist working in the 1960s he had considered a "sudden" extinction to be one that occurred over a few million years (quoted in Glen 1994, p. 41). The impact theory proposed that extinctions would take place in weeks or months. Since the resolution achieved by stratigraphy is imprecise, earth scientists customarily expected that stratigraphic horizon markers would vary considerably from one location to another and there would be considerable vagueness in the scheme for mapping such horizons (p. 77).

In contrast, the impact theory proposed that the bolide crash would blast vast amounts of debris into ballistic trajectory and coat the earth with a thin layer of distinctive sediment (Alvarez et al. 1980). Since this layer would be rich in iridium and other siderophile elements, which otherwise are extremely rare in the earth's crust, it would form a worldwide pencil-thin, fingerprint-quality horizon marker. Therefore, according to Glen, converts to the impact theory would have to reject some long-standing and deeply entrenched beliefs (Glen 1994, p. 79).[4]

In a few cases, scientists who switched to the impact theory did so after conducting field observations at localities where the K/T horizon is particularly well exposed. Others switched when new evidence, such as the discovery of "shocked" minerals at the K/T extinction horizon, was regarded as particularly compelling. Glen notes, however, that the influence of particular data or research results on theory choice is usually unclear (p. 74). At any rate, says Glen, the net result of his case studies of "conversion" to the impact theory is that strong support is given to Kuhnian tenets: "Many aspects of this study . . . support Kuhn's view that a paradigm shift—*sensu latu*, the switch to the alternative hypothesis—carries with it, for the individual, a fundamental change in cognitive values. Indeed, the term 'cognitive value' seems inadequate in conveying the sense of profound change apparent in the entire range of perception and response following a conversion experience" (p. 79).

Glen criticizes Rachel and Larry Laudan for stigmatizing Kuhn's talk of "gestalt switches" and "conversion experiences" as "psychobabble" (ibid.; see Laudan and Laudan 1989). He asserts that these are genuine if poorly understood effects of radical theory change. He agrees with Paul Feyerabend that "bandwagon effects" play a large part in leading scientists to switch allegiance to a new theory (p. 79).

That converts to new theories undergo radical conceptual and perceptual change is supported by Glen's claim that a mass extinction hypothesis, once adopted, became by far the best predictor of which standards of theory choice would be accepted (pp. 68–69). Indeed, in the debate between volcanists and impactors, Glen asserts that opposing standards of theory choice were often so antinomic that mutual blindness and incomprehension resulted: "Standards of appraisal were the most frequent targets of attack in the debates. . . . It was as if the impactors were blind to even the strongest elements in the cognitive framework of the volcanists and vice versa—Kuhn's term, 'incommensurability of viewpoints,' is not adequate to the descriptive task here. Standards of all sorts floated in and out of question and power" (p. 69). Glen reiterates this conclusion at the end of his chapter: "The gestalts (mindsets) or cognitive frames of members of the opposing theoretical camps seemingly precluded mutually congruent viewpoints on any of the important debated issues, or even the assessment of evidence. So deaf did they appear to each other's arguments that Kuhn's view that such adversaries suffer an 'incommensurability of viewpoints' seems understated" (p. 89). It follows that on Glen's view the convert to the impact theory would change not only his factual beliefs but also his standards of theory choice and indeed his whole perceptual/conceptual/cognitive framework.

Presuming that Glen has accurately described the behavior of scientists in the instances of theory change arising within the mass extinction debates, what lessons should we draw? Two sorts of lessons seem possible, a boring one and an exciting one (the "boring" and "exciting" terminology comes from Newton-Smith 1981, p. 103). The boring lesson would be that, sadly, scientists often fail to live up to scientific ideals. Instead of changing theories only after a judicious weighing of the evidence in the light of accepted canons of deductive and inductive inference, scientists often jump on bandwagons or have been swept off their feet by "conversion" experiences. Such behavior might still shock some, but it need not discredit scientific ideals any more than the behavior of Renaissance popes invalidated the Ten Commandments. An is/ought distinction between the ideals of science and the actual, often deplorable, behavior of flesh-and-blood scientists can remain firmly in place.

The exciting lesson that might be drawn from Glen's data is that the is/ought distinction should be erased, and we should see that scientists' *actual* behavior on occasions of theory change is how they *must* behave. On this view, all of the criteria relevant to theory choice are rigidly embedded within particular theories so that to accept a new theory is ipso facto to accept a complete set of new theory-choice criteria. When you discard an old theory, you also repudiate the entire set of methods, standards, and values that supported that theory. In such a case, a shared basis for the rational assessment of theories is ruled out; "conversion" becomes the *only* way to shift theoretical allegiance. This construal of Glen's data thus bears radical epistemological consequences.

Kuhn's early writings are usually interpreted as drawing the exciting sort of lesson from the kinds of data adduced by Glen (see Newton-Smith 1981, p. 103). What about Glen? Does he derive radical epistemological consequences from his sociological investigations? His endorsement of Kuhn's description of opposing theories as "incommensurable" indicates that he follows the early Kuhn in deriving such conclusions (Glen 1994, p. 89). To say that theories are "incommensurable" does not mean merely that proponents of opposing theories are myopic, cantankerous, or dogmatic. "Incommensurability" means much more than this, but precisely what?

Newton-Smith (1981) distinguishes three senses of "incommensurability." The first is *value incommensurability* (p. 149). He identifies scope, simplicity, accuracy, consistency, and fruitfulness as the five basic values that all good theories will manifest (pp. 112–13). However, such values do not characterize every good theory to precisely the same degree or in exactly the same way. Theory T might exceed competing theory $T^*$ in accuracy but fall below it in scope. Scientist Smith might value accuracy far more than scope and so, other things being equal, would favor T over $T^*$. Scientist Jones, who values scope more than accuracy, makes the reverse judgment. If no rational means exist for Smith and Jones to reconcile their disagreement over values, value incommensurabilty occurs.

However, scientists might agree on their basic values but still have rationally irreconcilable disagreements over other standards of theory choice. For instance, two scientists might fully agree that good theories should have a broad explanatory scope, but disagree on the nature of scientific explanation (p. 111). If consensus on the nature of scientific explanation cannot be rationally achieved, what Newton-Smith calls *standard incommensurability* results (p. 150). For Newton-Smith a "standard" is apparently any criterion of theory evaluation or comparison other than the five basic "values" identified above.

Finally, and most radically, theories might be incommensurable because they display *radical meaning variance* (RMV) (p. 154; see also Kordig 1971). Suppose that the terms in which theories are expressed have radically different meanings in different theories. Some writers (e.g., Feyerabend 1975) have suggested, for instance, that "mass" means something entirely different in Einstein's theories than in Newton's. If this radical meaning variance extends to observational as well as theoretical terms, as RMV asserts, then theories cannot be rationally compared with respect to their observational basis. In this case, advocates of rival theories are, in a very real sense, speaking different languages that are not intertranslatable (Lambert and Brittan 1987, pp. 134–35). Thus, radical meaning variance means that, rather than expressing rational disagreement, proponents of opposing theories are simply talking past one another.

Glen does not indicate which, if any, of the above senses of "incommensurability" he intends. However, he does endorse Kuhn's phrasing; he even says that the term "incommensurable" is inadequate to express the actual degree of mutual incomprehension found between the contending parties (Glen 1994, p. 89). Further, Glen hints at radical epistemological consequences when he tells us that "converts" to a new theory experience a "profound" change in their whole range of perceptions and cognitive values (p. 79). Finally, the very subtitle of his book, "How Science Works in a Crisis," reflects the Kuhnian theme that science enters periods of "crisis" between eras of "normal" science. During such a "crisis" contending parties wield radically different standards and evince mutual incomprehension (Kuhn 1970). Glen apparently identifies the mass-extinction debates as just such a "crisis." It therefore seems fair to interpret Glen as endorsing the "exciting" views attributed to the early Kuhn. In the rest of this chapter I shall call the "Glen-Kuhn" view the view that competing theories are incommensurable in one or more of the above senses and therefore that the *only* way for scientists to switch theories is to undergo a "conversion" of the sort Glen describes.

The Glen-Kuhn view has dire consequences: Proponents of opposing theories are (unavoidably) reduced to shrill assertion, perhaps aided by employment of rhetorical tricks, appeals to authority, and ad hominem abuse. Similarly, the individual convert to a new theory would, on this view, be motivated by irrational—or at least nonrational—factors rather than persuaded by rational reasons. If scientific change, whether on the part of an individual or a community, thus cannot be explicated rationally, the only remaining choice is explanation in terms of prejudice, ideology, sociological factors, etc.

The above alleged consequences of the Kuhnian view were gleefully accepted and widely promulgated by Paul Feyerabend (1975, 1978). On the other hand, Kuhn himself emphatically denied that his aim was to make theory choice irrational or rule out shared criteria for the evaluation of theories (Kuhn 1977). However, in his effort to repudiate the apparently radical implications of his views, Kuhn substantially modified his original language (see Kuhn's postscript in Kuhn 1970; the book was first published in 1962). Some commentators say that Kuhn so qualified his original, apparently radical claims, that his later views are uncontroversial to the point of triviality (Toulmin 1972; but see Bernstein 1983 for a more charitable interpretation).

Whatever Kuhn's later aims, the damage, so to speak, had been done. It was the clarion call of Kuhn the revolutionary that was heard, not the muted tones of Kuhn the Social Democrat (Newton-Smith titles his chapter on Kuhn "From Revolutionary to Social Democrat"). For instance, Barry Barnes, one of the foremost recent sociologists of science, writes enthusiastically that Kuhn debunked the pervasive "myth of rationalism" (Barnes 1985, p. 98). The "myth of rationalism" is partially characterized by Barnes as follows:

> Rationalist doctrines emphasize the power of the reasoning capabilities which all individuals possess. Rationalist accounts of science see its growth as the product of individual acts of reasoning. By logical inference on the basis of their experience, individual scientists contribute to scientific progress, to the cumulative development of scientific knowledge and its gradually increasing correspondence with the reality it describes. (pp. 85–86)

According to Barnes, Kuhn undermined this whole view of science: "It will be clear by now that Kuhn's work uncompromisingly and comprehensively undermines the rationalist account of science. It repudiates the power of an autonomous individual reason; it rejects an individualistic account of research; and it denies that scientific change is a gradual evolutionary progression" (p. 94). Kuhn allegedly showed these things in large part through his arguments on the revolutionary nature of theory change (pp. 91–93). According to Barnes, Kuhn argues that there can be no shared standards to permit a common measure of the merits of opposed theories; that is, competing paradigms evince standard incommensurability (p. 93). For Barnes, as a sociologist of knowledge, the bonus here is that the radical Kuhn has apparently taken the study of science away from rationalist philosophers and given it to the sociologists of knowledge.[5] If no standards guide theory choice, it follows that it must depend on the (nonrational) collective decisions of scientific communities. The analysis of collective human actions and their motivations falls within the province of sociology (see Laudan 1977, pp. 196–222, for a good riposte to this attempted usurpation).

In fact, Kuhn did not reject all standards of theory choice. He proposes a *sort* of standard: New theories must solve the "anomalies" of the old theory and usher in a new period of puzzle-solving "normal" science (Kuhn 1970). The really interesting question is whether Kuhn really did show that it is *impossible* for theory change to proceed in the purportedly mythical rational manner that, according to Barnes, he has definitively debunked. Again, for their claim to be at all interesting, would-be debunkers of rationalism must show more than that scientists often stray from rationalist ideals. They must show that it is *in principle* impossible for theory change to proceed in the traditionally rational way.

A single counterexample is sufficient to confute such a claim of in-principle impossibility, so let us turn to the case of David Raup's "conversion" to the bolide impact theory of mass extinction to see whether it conforms to the Glen-Kuhn account of theory change or whether it is better explicated in terms of the old-fashioned rationalist canons of theory choice that Barnes says are now passé. I shall argue that after his "conversion" to the new theory Raup continues to go about his science in much the same way, accepts many of the standards of theory choice he always did, endorses many of the same old meth-

## The "Conversion" of David Raup

ods, and in general has not changed his whole worldview. Further, at no point was he confronted with value, standard, or meaning incommensurability. In short, I contend that in Raup's case we find no conceptual hiatus, no radical displacement of a whole constellation of concepts, values, standards, or meanings as expected on the Glen-Kuhn account.

On the contrary, I argue that Raup changed his mind in a traditionally rational manner in accordance with *deeply grounded* and *broadly shared* reasons. By "traditionally rational," I mean that Raup was rationally persuaded to change theories by evidence and argument based on those sorts of broadly shared standards, criteria, methods, techniques, data, and so on, that are traditionally recognized as providing *reasons* for belief change. By "deeply grounded" I mean that such standards, methods, and the like are not rigidly embedded in, and thus only deployable within, particular theoretical frameworks or narrow disciplinary specializations. Rather, their justification appeals to principles of rationality that cut across theories and disciplines—and "paradigms." For instance, Raup employed mathematical and statistical techniques widely employed in the sciences. Further, the standards guiding his inferences were not unique to a given theory but recognized across theoretical and disciplinary boundaries. It follows that the standards on which Raup's theory choices rest are *much* more deeply grounded and broadly based, and thus more rational, than proponents of the Glen-Kuhn view will admit.

Raup's considerable reputation rests upon theoretical studies conducted with a high degree of rigor and mathematical sophistication. Throughout his career he has shown exceptional openness to new ideas. In fact, he toyed with an impact model of mass extinction well prior to the Alvarez team's paper (Raup 1986, pp. 43–45). In the 1970s Raup performed computer simulations of localized comet strikes in areas such as Australia or New Guinea to see if these might bring about mass extinctions of endemic species. Raup's computer simulation "bombed" the earth's surface with these strikes. The results were negative; the simulations indicated that mass extinctions could not be achieved by random, or even directed, comet strikes. Raup remarks that his colleagues thought the whole project "crazy" and that he refrained from publishing his results from fear of ridicule (pp. 44–45). Though his results were negative, Raup implies that, unlike his colleagues, he did not reject impact hypotheses a priori.

## Drawing Out Leviathan

In 1980 *Science* asked Raup to review for publication the paper by the Alvarez group. He records his initial reaction to the manuscript:

> In my review of the manuscript, I found fault with certain aspects and made suggestions for elaboration of the research and for revising of presentation.... I ended my review, however, with a fairly unkind statement saying: "If a graduate student gave me this manuscript to read, I would see it as a brilliant piece (indicating that the student had enormous potential) but I would give it back to be done right!" I thought the work was sloppy and incomplete, however brilliant it may have been. (pp. 67–69)

Raup also records that he was irritated that the paper was much longer than the usual *Science* article and that he considered its style "pretentious" (p. 67). He immediately reflects on the motivation of his initial reaction:

> I have related some of my reactions to the Alvarez paper because I think they say something about the way science works. Deep down, being very uncomfortable with the iridium anomaly and its extraterrestrial interpretation, I struck out at it in some possibly irrational ways. The manuscript did need work when I saw it, but that's what the review process is for. The paper in its published version had many of the holes filled—holes that I, and surely other reviewers, had noted. The paper turned out to be a clear first statement of a hypothesis and the evidence for it. (p. 69)

Raup thinks that much of the initial hostility (presumably including his own) toward the impact hypothesis was due to a deeply ingrained professional bias against "catastrophism" (pp. 29–34). Catastrophism and its opposite, "uniformitarianism," are somewhat elusive terms. Raup cites the influence of Charles Lyell as the reason for the strongly uniformitarian bias of the earth sciences (p. 32).

Martin Rudwick notes that "uniformity" (the term "uniformitarianism" itself was coined by William Whewell) had a variety of different meanings for Lyell (Rudwick 1976, p. 188). In fact, the interpretation of Lyell on "uniformity" is still an issue in the professional literature. According to Richard Huggett (1997), Lyell adopted four uniformities:

> The uniformity of law, the uniformity of process (actualism), the uniformity of rate (gradualism), and the uniformity of state (steady statism). The first two are usually regarded as procedural rules prac-

tised by all geoscientists; the last two are substantive claims about the empirical world. Lyell's system of strict uniformitarianism was founded upon these four assumptions. The first assumption, and to a lesser degree the second, he shared with catastrophists; the third and fourth were the mainstays of his particular vision of the world. (p. 37)

James A. Secord agrees that Lyell insisted upon uniformity of law, kind, and degree, but insists that none of these proposals were substantive claims about the world; all were strictly regulative or methodological, intended by Lyell to place geology on secure inductive grounds (Secord 1997, p. xix). In other words, Lyell required geologists to explain earth history in terms of recognized physical law and not to invent new causal processes or postulate of known processes that they operated with a degree or intensity never observed. Similarly, Rudwick holds that Lyell's "actualism" was the two-pronged methodological rule that causes assigned to geological phenomena should not differ in kind *or* intensity from those causes previously observed (Rudwick 1990, pp. ix–x). Rudwick claims that Lyell's actualism reflects John Herschel's doctrine that scientific explanations should invoke only *verae causae,* "true causes" that could be *observed* to produce the relevant effects (p. ix; Herschel 1831). "Actualism," so defined, rules out any explanation invoking agents never shown to produce the given sort of effect or agents postulated to produce their effects with a never-observed degree or intensity.

Roy S. Porter, on the other hand, in the *Dictionary of the History of Science,* restricts "actualism" to the claim that geological phenomena should be explained in terms of observable processes (Porter 1981, p. 5). He construes uniformitarianism as a stricter doctrine, which further stipulates that those processes should be postulated to have always acted at about their present level of intensity (p. 430). So Porter, like Huggett and unlike Rudwick, terms "actualism" the methodological rule that geological hypotheses postulate only observed processes, but does not impose the further requirement that those processes always operated with the same degree or intensity.

To avoid confusion, I shall follow Porter in regarding actualism as the methodological requirement that geological hypotheses postulate only presently observable causal processes. Uniformitarianism includes actualism, so defined, and further stipulates that geological processes be assumed to have always operated at about their present rate or intensity. I shall call "catastrophist" any geological theory postulating a causal process of unobserved type *or* intensity. It follows that, on my definition, a theory can be both catastrophist and actualist (though not both catastrophist and uniformitarian).

Returning from tedious semantic matters, by 1980, when the Alvarez paper appeared, humans had observed impacts by comets and large meteorites. Perhaps the most dramatic of these was the Tunguska event of 1908, usually interpreted as the impact of a small comet fragment (Sagan 1980). However, none of these impacts was nearly large enough to cause mass extinctions. Known earthly impact structures, such as Meteor Crater in Arizona, were also far too small to have been caused by agents of mass extinction. Really big impacts were believed to have been limited to the very early era before the earth's present surface had formed (Raup 1991, p. 157). Of course, in 1980 it was recognized that the uniform operation of known agents could produce local catastrophes, as was vividly demonstrated by the eruption of Mount St. Helens in May of that year. Until July 1994, when comet Shoemaker-Levy 9 struck Jupiter, humans had never observed, on this planet or any other, an impact of sufficient magnitude to threaten mass extinctions. Though humans had observed destructive events of the general *kind* hypothesized by the Alvarez hypothesis, they had never witnessed anything like one of the requisite *intensity.*

However, by 1980 earth scientists had abandoned strict uniformitarianism, if they had really ever adhered to it (though they have remained actualist). It had already been recognized that some geological causes far exceeded in intensity anything humans had witnessed. For instance, J. Harlen Bretz (1923a, b), though he faced prejudice initially, eventually persuaded his colleagues that the Channeled Scablands in Washington State had been shaped by catastrophic flooding of an enormous scale. Similarly, geologists had long accepted the existence of ice ages, which postulated a degree of continental glaciation never observed in historic times. Raup may therefore have overestimated the anti-catastrophist biases of his colleagues.[6] His own attitudes, as evidenced by his willingness to test impact scenarios as early as the 1970s, do not appear as closed-minded as he seems to recollect. There is inevitably much uncertainty here since even severely honest introspection is questionably reliable when it comes to revealing one's deepest motives and attitudes. *Remembering* such motives and attitudes from several years' distance is even less reliable.

## The "Conversion" of David Raup

The trickiness of memory is shown by Raup's recollection of his early reaction to the impact debates that arose soon after the publication of the Alvarez team's paper:

> Where did I stand on these arguments at the time? The opinion of one person, not an active participant, is not important, but I can be somewhat more accurate (perhaps) describing my own reactions than trying to divine what other people were thinking several years ago. My recollection of the time is that I was cautiously optimistic about the impact hypothesis. When asked by friends and colleagues, I usually said something like "I sure hope they are right but the hypothesis has some very serious problems." (Raup 1986, p. 71)

This recollection is at odds with his assertion, recorded two pages earlier in his book, that "deep down" he had been "very uncomfortable" with the iridium anomaly and its extraterrestrial interpretation when he first read the Alvarez paper (p. 69). Perhaps some holes in the original manuscript had been plugged, but it still seems odd that, within a year or so of the paper's publication, Raup would have overcome what he recalls as major philosophical reservations about the extraterrestrial hypothesis. Perhaps, then, his attitude toward the impact hypothesis had all along been somewhat friendlier than he later records.

Raup's active participation in the extinction debates began with his collaboration with J. John (Jack) Sepkoski Jr., also a University of Chicago paleontologist. Early in 1983 Raup and Sepkoski began a rigorous statistical examination of extensive data that Sepkoski had compiled in his *Compendium of Fossil Marine Families* (1982). This compendium lists the first and last known occurrences of species of over thirty-five hundred marine fossil families.

After several months of number crunching, a very surprising result seemed to be emerging. Regularly spaced episodes of mass extinction seemed to emerge from the data (Raup 1986, pp. 115–22). Raup, a very capable statistician, was of course aware that random distribution can mimic periodicity, but even strict statistical tests, conducted at a 99.9 percent confidence level, failed to eliminate the periodicity (p. 122). Raup and Sepkoski concluded that episodes of mass extinction were distributed nonrandomly and had a periodicity of 26 million years. This was not the first time periodic mass extinctions had been proposed, but, Raup implies, it was the first time that such

David Raup.
Photograph © 1996 by Judith T. Yamamoto.

a hypothesis had stood up when tested against hard historical data (pp. 107–108). What was still needed was a plausible mechanism that would produce periodic pulses of mass extinction.

At this point a number of astrophysicists jumped in to offer theories that would account for the perceived periodicity of mass extinction; the April 19, 1984, issue of *Nature* had four such articles. The explanation that developed as the leading contender was the one that came to be named the "Nemesis" hypothesis. This hypothesis proposed that the sun has a small, dark companion star in a highly eccentric orbit that brings it close to the sun every 26 million years (Whitmire and Jackson 1984; Davis, Hut, and Muller 1984). When approaching the sun, the companion star passes through the Oort cloud, a halo of

## The "Conversion" of David Raup

billions of comets orbiting the sun beyond the outermost planets. The gravitational disturbance caused by the near passage of the companion star would send millions of comets cascading into the inner solar system; some of these would almost certainly strike earth. The imagined devastation of these comet strikes, held to be sufficient to cause mass extinctions, led the Davis team to name the companion star "Nemesis."

The Nemesis hypothesis fanned the flames of the already heated controversy over mass extinction. Willingly or not, Raup was pushed to the forefront as a champion of the impactors. He capably defended his claims against a number of critics (Raup 1986, pp. 151–55). From a distance of several years it is now safe to say that the Nemesis hypothesis was not an overall success. Critics such as Antoni Hoffman have persuasively argued that the alleged periodicity of 26 million years is a statistical artifact and that Sepkoski's data are in fact compatible with a random walk (Hoffman 1989a). In his 1991 book *Extinction: Bad Genes or Bad Luck?* Raup reports that further statistical tests of the Sepkoski data by other scientists and statisticians have produced decidedly mixed results (Raup 1991, pp. 164–65). He notes also that most astronomers have rejected the idea of a dim stellar companion to the sun (p. 165).

Raup concludes *The Nemesis Affair* with an interesting chapter entitled "Belief Systems in Science." He concedes that new hypotheses must bear the burden of proof, but he argues that when a hypothesis runs contrary to strongly embedded philosophical prejudice, it cannot get a fair hearing (Raup 1986, p. 196). He offers a judicial analogy. In a particular court case the plaintiff sued for compensation on the grounds that his prayers had caused the defendant to win the lottery (pp. 197–201). The judge found against the plaintiff on the grounds that the efficacy of prayer cannot, in principle, be demonstrated scientifically.

Raup counters that the efficacy of prayers, chants, incantations, etc., is plainly just as testable as any scientific proposition (p. 200). This is a dubious claim, but I shall accept it for the sake of argument. Raup concludes that in reality the judge simply considered the plaintiff's case to be based on a hypothesis not worth testing (ibid.). He implies that philosophical prejudice likewise excludes many promising scientific theories from due consideration.

Is it a matter of sheer prejudice whenever philosophical considerations exclude a hypothesis from being tested? It is a simple fact that the human imagination can produce many more hypotheses than can possibly be tested. Thus, it is unavoidable that many conceivable hypotheses will have to be ruled out on philosophical grounds (see Rothbart 1990 for a very astute discussion of pre-test criteria). Certainly there have been many occasions on which such grounds were wrongly applied and promising new hypotheses were unfairly squelched. However, to claim that this was so in any particular case requires much detailed argument. More importantly, the impact hypothesis certainly did *not* suffer such a priori exclusion. It *was* tested early and extensively.

Perhaps Raup intends only to make the weaker charge that uniformitarian prejudices prevented his colleagues from seeing just how strong the evidence for the impact hypothesis actually was. To argue this in a non-question-begging way would require reentry into the ongoing debate between impactors and gradualists. Besides, it still needs to be shown that such "philosophical prejudices" as a commitment to actualism or uniformitarianism are avoidable within scientific disciplines. In fact, as Kuhn's own analysis of "normal science" stressed, any scientific discipline will have to be guided by certain deeply embedded philosophical commitments.

Does Raup's account of his own "conversion" lend support to the Glen-Kuhn analysis? Raup implies that he underwent a profound philosophical change when he accepted the catastrophic impact theory. However, Raup may have been mistaken about the nature of his own and his colleagues' philosophical commitments. As noted earlier, by 1980 there were probably few strict uniformitarians left in the earth sciences. It is fair to say that earth scientists were generally actualists, as I have defined the term; that is, they sought to explain earth history in terms of known causes, but were tolerant of explanations proposing processes of unobserved degree. Of course, the Alvarez hypothesis postulated a bolide impact of *far* greater destructiveness than had been observed, so that tolerance may have been stretched to the limit.

Raup's early willingness to examine impact scenarios did show that he was more open to such hypotheses than were most of his colleagues. He admits that he was initially somewhat receptive to the impact hypothesis, but he still feels that his first reaction to the Alvarez team paper was prejudiced:

## The "Conversion" of David Raup

But as I look back over that review, I see that I was challenging things in the Alvarez paper that I would now call minor misdemeanors, if that. In spite of my preconditioning to be receptive to impacts as a cause of extinction, I was finding fault in what was pretty surely a classic reactionary mode. If I were to receive the same paper now, describing perhaps an iridium anomaly at some other point in the geologic record, I am reasonably sure I would accept most of the things I found fault with in 1980, because I am now a "believer" in large-body impacts and their iridium signature. (Raup 1986, pp. 205–206)

This last bit of testimony fits rather well with the Glen-Kuhn account. Raup seems to be saying that his acceptance of the impact hypothesis has substantially altered his standards for the evaluation of evidence. He is not saying that he once had legitimate objections that have now been answered. Rather, he is asserting that in his present judgment his original objections were expressions of prejudice, prejudice due to the theoretical blinders he wore at the time. I do not know whether Raup has been influenced by Kuhn; he seems to be self-consciously describing himself in Kuhnian terms. Glen could easily interpret such testimony as evidence that Raup, having undergone a "conversion" to the impact theory, has concomitantly adopted a set of theory-choice criteria so radically different that his old objections now seem like mere prejudice.

When we look at the actual objections that Raup made in his review of the Alvarez paper, things are not so clear. His first objection was that the Alvarez team had not adequately examined iridium distributions elsewhere than in the K/T boundary site they had studied (reprinted in Raup 1986, p. 68). Does Raup mean to imply that he would now regard such an omission as a mere peccadillo, at worst? The whole argument of the Alvarez team rested upon the claim that iridium anomalies were present only in the stratigraphic markers of large extraterrestrial impacts. High concentrations of iridium in strata clearly uncorrelated with impacts or mass extinctions would count strongly against that hypothesis. Indeed, Raup is simply expressing Mill's methods of Agreement and Difference: Iridium anomalies, the supposed effect, should be absent when the purported cause is absent. Thus, Raup's criticism still seems a powerful one and in no sense a captious or niggling objection. One is left wondering just what to make of Raup's mea culpa.[7]

## Drawing Out Leviathan

At this point it would be valuable to go over all of Raup's objections to the Alvarez paper in his review for *Science*. Unfortunately, except for the first page, which is reprinted on p. 68 of *The Nemesis Affair*, this document is not available (however, note personal communications from Raup in the notes to this chapter). If we accept Raup's testimony, his chief initial objections to the paper were that it was "sloppy and incomplete," not that it was fundamentally wrongheaded or otherwise grossly defective (Raup 1986, p. 69). On the first page of his review, Raup notes the great potential significance of the paper and states the aim of his critical comments: "My comments are directed towards encouraging the authors to fill in some gaps in their reasoning and to streamline the paper so that it is the cleanest and least ambiguous presentation possible" (p. 68). These remarks evince a helpful, even benign attitude toward the paper and its hypothesis. The only "unkind" comment Raup records having made was his comparison of the Alvarez team's paper to the work of a brilliant graduate student (p. 69).[8]

Nothing in this indicates that Raup's attitude toward the Alvarez paper had changed from uncomprehending hostility to a convert's enthusiastic acceptance. The main change in attitude seems to be that the alleged sloppiness and incompleteness of the paper was initially judged severe enough to merit a delay of publication and was later judged not to. Like St. Augustine's *Confessions*, *The Nemesis Affair* engages in self-reproach that is out of proportion to the sins actually committed.

More fundamentally, nothing in Raup's testimony gives the least indication that his earlier views were in any sense incommensurable with his later ones. Raup does not indicate an upheaval in his fundamental scientific values or that he had been forced to adopt entirely new standards. He certainly does not indicate that the meanings of theoretical and observational terms had changed radically for him.

What about the "K/T boundary" though? Raup does not specifically discuss the change in the meaning of this term, but, as indicated earlier, would not "K/T boundary" mean something radically different to the proponent of the impact hypothesis? Previously it had meant a rather vague stratigraphic zone; on the Alvarez view it is a pencil-thin, fingerprint-quality horizon marker. Is this not an instance of radical meaning variance (RMV)?[9]

Recall that for RMV to hold, *observational* terms must mean some-

## The "Conversion" of David Raup

thing entirely different on the new theory. The observations on which the Alvarez team based a new understanding of the K/T boundary involved the measurement of anomalously high concentrations of iridium in a thin layer of clay. "Iridium," "clay," "layer," and various instrument names would seem to mean pretty much the same for the Alvarez team as for any other scientists. The data supporting the new understanding of the K/T boundary seem expressible in entirely neutral terms. Proponents of the Glen-Kuhn view will have to look elsewhere for an instance of RMVs.

The upshot is that Raup's testimony in *The Nemesis Affair* does not support the Glen-Kuhn view that acceptance of the impact hypothesis would entail a conceptual "conversion" or "gestalt switch." Raup apparently accepted the radical new theory without undergoing any sort of catastrophic cognitive upheaval. Even in *The Nemesis Affair* his commitment to the impact view is expressed in a cautious and qualified way: "My hunch at the moment is that the evidence for meteorite impact at the K-T boundary is much stronger [than the evidence for massive volcanism], but the returns are not all in" (Raup 1986, p. 93). This is hardly the zealous endorsement one would expect from a convert.

The full extent of Raup's "conversion" thus seems to be a shift in attitude from open-minded skepticism to cautious acceptance. This change of attitude seems perfectly explicable in the old-fashioned rationalist terms derided by Barnes. Raup indicates that it was the rigorous statistical tests he and Sepkoski performed on Sepkoski's data that made the difference between their theory of periodic extinction and earlier ones (Raup 1986, p. 108). Raup does record one incident which would no doubt be seized upon by advocates of the Glen-Kuhn view. He records an occasion on which he and Sepkoski stood on the other side of the room looking at a computer printout of Sepkoski's *Compendium* data looking for, as Raup puts it, "a gestalt that would lead us in interesting directions" (p. 115). However, this incident occurred at a very preliminary stage, well before the extensive and thorough statistical examination of the data. Like any good scientist, Raup no doubt realized that all sorts of "gestalts" can arise from a preliminary look at the data; only rigorous testing can confirm or deny such initial impressions. In fact, Raup is at pains to explain to the reader just how random distributions can create the impression of periodicity (pp. 116–20).

Testimony, no matter how scrupulously honest, is far from perfectly reliable. Therefore, let us turn to some of Raup's published writings in the period before, during, and after his acceptance of the bolide impact hypothesis. Careful examination of key publications during these periods should reveal the changes, if any, in his standards of theory choice, methodologies, worldview, etc.

As implied earlier, the most shocking aspect of the bolide hypothesis was the fact that the Alvarez team's paper implied a literally almost instantaneous mass-extinction event. Paleontologists had long been accustomed to thinking of mass extinctions as being "sudden" only in the geological sense, that is, as occurring over periods extending to a few million years. Of course, Darwin himself was aware that certain extinctions were apparently abrupt, as when he commented on the "wonderful suddenness" of the disappearance of the ammonites at the close of the Secondary Period (Darwin 1979, p. 322). However, Darwin viewed the appearance of suddenness as an artifact of the geological record: Geological "simultaneity" need not relate to the same thousandth or even hundred thousandth year (p. 326).

In general, Darwin did not consider that extinction was a phenomenon requiring any special explanation; natural selection accounts for extinction just as it explains adaptation. He argued that the size of populations of organisms is always held in check by various "injurious agencies," especially competition with other forms of life. These agencies are often severe enough to cause the rarity and ultimately the extinction of a species (p. 323).

Darwin did not attempt to explain mass extinctions because he did not recognize such a phenomenon. He explained only what today would be called "background" extinctions, the gradual, ongoing extinction processes that are due to the continuous operation of natural selection. Background extinction is piecemeal—a species here, a species there—and scattered randomly through geological time. Mass extinction, on the other hand, is ecologically pervasive, cutting across taxonomic and ecological boundaries in a limited amount of geological time.

Once the reality of mass extinctions in the geological record had been recognized, it was possible to formulate theories of mass extinction compatible with Darwinism. A Darwinian view of mass extinction could view a very gradual and widespread worsening of environmental conditions, such as a slow cooling of the global climate, as

## The "Conversion" of David Raup

multiplying and exacerbating the "injurious agencies," leading finally to widespread extinction. (See Stanley 1987 for a treatment of mass extinction very much along these lines.) The generous stretches of time allowed within a geological instant could easily accommodate such a Darwinian view of apparently sudden mass extinction.

In his *Extinction: Bad Genes or Bad Luck?* Raup argues that the bolide impact theory entails a very different view of mass extinction. On his view, the devastation caused by the bolide impact would be so sudden and so catastrophic that what he calls "wanton extinction" would prevail (Raup 1991, pp. 185–90). Wanton extinction is selective, but not in the Darwinian sense. When environmental conditions are suddenly (truly suddenly) made catastrophically worse, those few organisms that survive will tend to do so because, by chance, they were preadapted to these uniquely extreme conditions. For instance, it has been conjectured that cockroaches would be among the most numerous survivors of a nuclear holocaust since they just happen to be highly resistant to radiation. Presumably, high levels of radiation were not among the selective pressures shaping cockroach evolution. Therefore, Darwinian fitness does not explain the high rate of survival of cockroaches in a high-radiation environment (p. 186).

Actually, Raup's presentation of the contrast between wanton selection and natural selection is somewhat misleading. The inapplicability of Darwinian fitness to the extinctions caused by a massive bolide impact has nothing to do with the suddenness or severity of the "injurious agency." As James Lennox puts it, "[Darwinian fitness] is a measure of the selective advantage of one genotype over another *within* a population. Unless there is differential reproduction within a population, there is no Darwinian selection. Thus, when entire genera or families are extinguished, this is *not* what biologists call natural selection" (Lennox, private communication). A massive bolide impact does not select the fittest members of populations; it wipes out whole populations. When whole genera and families are destroyed, as occurred in the mass extinctions, distinctions in Darwinian fitness between individual organisms is irrelevant.

Raup asks whether species most often go extinct because their members are less fit than their competitors ("bad genes," the allegedly Darwinian view) or because they are the unlucky victims of cataclysms that wipe out entire families or genera of organisms, irrespective of the Darwinian fitness of individuals. He thinks the latter sort of extinction has played the larger role in the history of life (Raup 1991, p. 191).

"Wanton extinction," which is implied by the bolide hypothesis, is truly at odds with the received and deeply entrenched Darwinian view of extinction. Is this the major conceptual shift that Raup had to undergo when he accepted the bolide theory? Raup coauthored with Steven Stanley a textbook on paleontology, *The Principles of Paleontology*. The treatment of mass extinction in this book is consistent with the Darwinian model; it entertains a variety of gradual mechanisms that could have led to a decrease in speciation or a slow worsening of "injurious agencies" (Raup and Stanley 1978, pp. 438–48). Did the impact hypothesis therefore engender a global shift in Raup's thinking about mass extinction—a shift from the conservative Darwinian view to the radical concept of wanton extinction?

An article first published in *American Scientist* in 1977, "Probabilistic Models in Evolutionary Paleobiology," shows that Raup's thinking about mass extinctions was rather subtle at this point. He begins the article by noting that major events in evolutionary history, including mass extinctions, are usually explained in terms of specific deterministic causes:

> Most interpretation of the fossil record has centered around finding specific causes for specific evolutionary events. Why was there a sudden diversification of life in the late Precambrian? Why did one group of corals replace another? Why did the dinosaurs go extinct? Why did the human species evolve when it did? And so on. The approach has been highly deterministic; each event has been treated as unique, and although generalizations have been made there have been few attempts to look at groups of events in a probabilistic way. (Raup 1981, p. 51)

For instance, in *Principles of Paleontology* Raup and Stanley suggest that regressions of shallow epicontinental seas and the interruption of warm currents may have resulted in considerable climatic cooling, and concomitant extinctions, at the end of the Cretaceous (Raup and Stanley 1978, p. 442).

To construct an alternative probabilistic model, Raup devised a computer simulation in which evolution was modeled as a purely random process: "The work uses computer simulation and Monte Carlo methods to generate imaginary evolutionary trees . . . a branching

## The "Conversion" of David Raup

pattern of lineages is built up using random numbers. In effect, we are asking: What would evolution have looked like if it were a completely random process—without natural selection and adaptation and without predictable environmental change" (Raup 1981, p. 54). Actually, his phrasing is quite misleading. Raup's probabilistic model did not assume that natural selection does not occur or that there are no monotonic trends of environmental change. Rather, the possibility that he considers is that, given the enormous expanse of evolutionary time, and given the complexity of the adaptive relationships between organisms and between organisms and their environment, natural selection might behave over the long run in a mathematically random manner: That is, this model does not deny causality in a particular selection event; it regards such causes as randomly distributed over geological time (p. 52). On this model, selection pressures will tend to cancel each other out in the long run and thus will tend to erase each other's effects. However, the stochastic nature of this process permits occasional violations of such homeostasis. Occasionally an irreversible event such as speciation or extinction ends the process, but these events will occur in a mathematically random manner through geological time.

The initial aim was for the random models to serve as a null hypothesis for comparison to the actual evolutionary record (p. 54). To his surprise Raup found that his stochastic models very closely resembled real-world phylogenies with branching and extinction patterns very much like those reconstructed from the fossil record. Raup even conjectures that some so-called mass extinctions might be no more than statistical effects: Given enough time, the geologically simultaneous, coincidental occurrence of numerous independent extinction events is to be expected. Still, some extinction events, such as the Cretaceous-Tertiary extinctions, were so massive that they cannot be explained away in this manner. For such events specific deterministic causes must be sought (p. 55).

The upshot is that Raup sees two possible ways of explaining the major events in evolutionary history: "The real world of the fossil record probably contains a mixture of two types of events (or sequences of events): those caused by specific, nonrecurring phenomena and those so complex and unpredictable that they are best treated probabilistically in groups" (p. 58).

Apparently, as of 1977 Raup was looking for a unique ("nonrecurring") deterministic phenomenon to explain such major events as the Cretaceous-Tertiary extinctions; such an extinction was just too massive to be merely the stochastic effect of independent random extinctions.[10] A massive impactor would fill the bill nicely. Here was a nonrecurring (in the original Alvarez formulation) phenomenon that could explain an extinction event of any size. Further, accepting such a cause for really major mass extinctions would not mean a wholesale abandonment of Darwin or natural selection, as Raup himself emphasizes (Raup 1991, p. 192). Natural selection had been accommodated in Raup's surprisingly successful stochastic models. Unless an extinction event was truly massive, too large to be explained as the coincidence of many independent random extinctions, no question of considering a non-Darwinian type of extinction need arise. Further, truly massive extinctions need only be interpreted as permitting, not requiring, hypotheses of non-Darwinian extinction. Depending on the evidence, agents of mass extinction friendlier to Darwinian tenets, like gradual global cooling, might still be considered.

Given the above circumstances, it appears that Raup's 1977 work prefigured his openness to the impact hypothesis. Here was a specific, deterministic agent of exceptional power. Invoking such a cause in the cases of the few truly massive extinctions would hardly be a wholesale abandonment of Darwin. Far from having to postulate a radical conceptual hiatus between Raup's 1977 work and the Alvarez hypothesis, the former would seem to have prepared the ground for reception of the latter.

The chief complication here is that the Alvarez hypothesis, in accordance with Raup's 1977 expectations, postulated a nonrecurring agent of extinction, but the actual impact hypothesis advocated by Raup proposed a recurring cause. Was the transition from looking for a nonrecurring cause to hypothesizing a recurring one thus the great conceptual leap we are searching for? To see, suppose we look for the specific reasons that Raup first sought an extraterrestrial explanation for mass extinctions. These are found in a paper Raup coauthored with Sepkoski and presented at a conference at the University of Northern Arizona in 1983 (Raup and Sepkoski 1986).

This paper, "Periodicity in Marine Extinction Events," indicates that it was the perceived twenty-six million year periodicity of mass extinction that led Raup and Sepkoski to the extraterrestrial hypotheses. No known earthly or solar periodic processes operate on the appropriate time scale; by default astronomical causes were sought (Raup and Sepkoski 1986, p. 24). Statistical tests of cratering periodicity were thought to show a strong correlation with the Raup-Sepkoski extinction periodicity (pp. 24–26). These results, and other recognized evidence for massive impacts, supported their effort to seek a causal mechanism in known or conjectured astronomical periodicities.

What, then, led Raup and Sepkoski to the view that extinctions are periodic? The reasons given are extensive mathematical tests of the data in Sepkoski's *Compendium*. Fourier analysis and nonparametric randomization tests were applied to the data, and periodicity was apparently confirmed with a very high degree of confidence (pp. 16–20). These seem to be recognized mathematical techniques applied widely within the sciences. At least, there is no indication that one must have already "converted" to the impact hypothesis before the application of these techniques will be viewed as cogent.[11] In fact, Raup and Sepkoski report that the Fourier program they used came from a book written by J. C. Davis, *Statistics and Data Analysis in Geology*, that was published in 1973, long before the impact hypothesis (p. 16). Overall, there is no indication here that the reasons for considering impact hypotheses required major conceptual or methodological shifts for Raup or Sepkoski; the decision appears largely to have been made by default—only a hypothesis of periodic impact seemed to fit their data and background knowledge.

Finally, let us turn to "The Case for Extraterrestrial Causes of Extinction," a paper written by Raup in 1989, some years after the Nemesis controversy had died down. Here Raup is largely concerned with reconciling impact hypotheses with stratigraphic evidence. Some impact hypotheses postulate massive cataclysms caused by single impacts at extinction horizons. However, the final occurrence of specimens in the fossil record usually indicates stepwise or gradual extinction for higher taxa; that is, the last occurrence of specimens of the component species are found at various levels below the horizon. If the extinctions of species within a higher taxon did occur in such a stepwise manner, this is inconsistent with the view that the taxon was eradicated by a single, instantaneous cataclysm. Of particular interest here is the fact that Raup calls for an approach to this problem using "new methods of analysis" (Raup 1989, p. 423).

Chiefly these new methods need to take account of what Raup calls the Signor-Lipps Effect after an important paper by P. W. Signor III and J. H. Lipps (1982). The Signor-Lipps Effect occurs when the vagaries of fossil preservation cause an extinction that was actually instantaneous to appear gradual or stepped. Raup therefore seeks new methods that would allow scientists to distinguish between extinctions that were actually gradual and those that only appear so (Raup 1989, p. 428). These methods will involve the development of mathematical techniques that can compare the fossil record with expectations generated by hypotheses of sudden and gradual extinction (pp. 428–30).

There is simply no indication that the new methods Raup adumbrates will rest upon mathematical principles or statistical practices of an entirely new sort: That is, though the methods may be new, the mathematical and statistical principles that justify their application need not be. Raup is here merely trying to extend his career-long aim to apply statistical methods to paleontological hypotheses.

In conclusion, a review of Raup's testimony in *The Nemesis Affair* and a sampling of his publications on extinction from periods before and after his "conversion" fail to offer any data in support of the Glen-Kuhn view. On the contrary, the plain appearance is that Raup came to accept the impact hypothesis on the basis of considerations of a sort that the traditional rationalist would call "reasons." There is also no indication whatsoever that Raup's change in views forced him to undergo a radical shift in his whole perceptual/conceptual/cognitive makeup. With respect to methods, standards, and general worldview, the post-impact Raup seems quite consistent with the pre-impact one.

Here I anticipate the following sort of objection: It may very well be that David Raup did not undergo a true conversion when he accepted the impact hypothesis. Perhaps it was not a conceptual upheaval for him; maybe Raup was epistemically preadapted to the catastrophic conceptual changes brought by the impact theory. But surely Raup was an innovator, a figure on the cutting edge—hardly representative of his discipline. What about those, surely the vast majority of his colleagues, who did not evince such a remarkable openness to new ideas as did Raup? Surely these more conservative colleagues were faced with a stark alternative: Remain a gradualist and cling to your old Darwinian values, or surrender those values in toto and convert to one of the two catastrophic hypotheses. Thus, the Glen-Kuhn analysis still applies to the vast majority of earth scientists, even if there are a few remarkable exceptions like Raup.

Though Raup was an innovator, he appealed largely to *conservative* standards. His *application* of statistical methods may have been new, but the methods themselves were well grounded in recognized mathematical principles and techniques. Innovative, even radical theories can appeal to conservative standards.

As a case in point, Glen himself makes the striking observation that the impact theory was supported by certain conservative standards and criteria:

> The impactors' evidentiary appeal, different in certain respects from that of the volcanists, has been mainly to *prevailing criteria* by which to judge the origin of their data. For example, the siderophiles (iridium, etc.) in the K/T boundary bed must have come from the sky, since no endogenous, earthly cause is known to concentrate them ubiquitously in sediments on the Earth's surface; global iridium distribution at the same horizon could only be accomplished through ballistic transport unique to a bolide impact; and shocked quartz with multiple sets of shock lamellae could only have been produced by meteorite impact or an atomic-bomb explosion. . . . The volcanists—in contrast to the impactors, who, *notwithstanding their radical hypothesis, have argued mainly from conventional standards*—have focused instead on the vagaries and suppositions that lie hidden in and beyond the standards of appraisal and orthodox knowledge. (Glen 1994, p. 61, emphasis added)

The fact that the impactors appealed mainly to conventional standards is completely at odds with the claim, which I have attributed to the radical Kuhn, that new theories incorporate new standards which must simply displace those associated with old theories.

Here Glen might charge that I have arrived at the above conclusion only by ignoring vast amounts of the data that he has adduced. According to Glen, the overwhelming majority of scientists proclaimed their loyalties for or against the impact hypothesis at the beginning of the controversy and thereafter haven't budged (pp. 76–77). Scientists divided up on the question very strictly along disciplinary or subdisciplinary lines (pp. 50–55). Such commitments pro or con then became by far the best predictors of what theory-choice criteria would be adopted by individual scientists in the debates (pp. 68–69). Finally, practically all scientists interviewed expressed strong opinions for or against the impact theory whether or not they had in-depth knowledge of the issues (p. 47). Does all of this not indicate that in fact the vast majority of scientists jumped to a conclusion about the impact hypothesis and have simply sought to rationalize that decision in the subsequent debates? Do these data not show that scientists made "subjective" choices and then deployed whatever rhetoric, standards, or evidence they thought most effective in defending their choices? How can these data be squared with any sort of rationalist model of theory change?

Let me emphasize that in this chapter I am not presenting or defending *any* "model of theory change": That is, I am not here advocating any model of how scientists *actually* behave when confronted with radical new theories. (In the next chapter I do make some claims about actual scientific behavior.) I have argued that David Raup did in fact behave in a conventionally rational manner, but I admit that he may be the exception in this regard. What I *am* defending is the claim that in situations of theory choice scientists typically have at their disposal a wide array of *broadly shared* and *deeply grounded* standards, criteria, methods, techniques, data, etc., and that these are sufficient to permit fully rational decisions about theory acceptance or rejection.[12]

In answer to the above anticipated objection from Glen: Yes, scientists in the mass extinction imbroglio may have acted the way you describe, but they did not *have* to. As I emphasized earlier, to say that opposing positions were "incommensurable" must amount to more than simply saying that the opposing parties were intransigent, myopic, or cantankerous. To be at all interesting it must assert that *in principle* mutually agreeable values, standards of theory choice, or shared concepts for describing data were unavailable. I think that my argument in this chapter has shown that this was decidedly *not* the case. To all appearances, David Raup came to accept the impact hypothesis in a manner at odds with the Glen-Kuhn view of "conversions" and in conformity with the "myth of rationalism" castigated by Barnes. Rational means of agreement were available, and some scientists, even cutting-edge theoreticians like David Raup, actually used them.

The argument of this chapter supports views reached by Stephen Toulmin in his *Human Understanding.* In a searching and highly corrosive examination of *The Structure of Scientific Revolutions,* Toulmin sets Kuhn the historian against Kuhn the philosophical sociologist:

## The "Conversion" of David Raup

> As his [Kuhn's] historical analysis makes clear, the so-called "Copernican Revolution" took a century and a half to complete, and was argued out every step of the way. The world-view that emerged at the other end of this debate had—it is true—little in common with earlier pre-Copernican conceptions. Yet, however radical the resulting change in physical and astronomical *ideas and theories,* it was the outcome of a continuing rational discussion and it implied no comparable break in the intellectual *methods* of physics and astronomy. If the men of the sixteenth and seventeenth centuries changed their minds about the structure of the planetary system, they were not forced, motivated, or cajoled into doing so; they were given reasons to do so. In a word, they did not have to be converted to Copernican astronomy; the arguments were there to convince them. (Toulmin 1972, p. 105; emphasis in original)

Toulmin's conclusions about the continuity of methods and the rationality of the debates, if applied to the mass-extinction controversy rather than the Copernican Revolution, serve as an excellent summary of what I have argued in this chapter.

We may conclude by drawing a general lesson: When looked at from a broad perspective, many scientific changes look revolutionary. By the time they end, the world has been turned upside down. When such "revolutionary" changes are examined closely and in detail, however, continuities abound. Methods, techniques, standards, values, and philosophical commitments are broadly shared by participants in the debates, and often remain fairly constant across conceptual changes. Proponents of Kuhn-style analyses have often claimed that their view is more consistent with the history of science. But when history is examined closely, "evolution" rather than "revolution" seems the better metaphor for scientific change.

# 4

# ARE DINOSAURS SOCIAL CONSTRUCTS?

Because professors are a touchy and querulous lot (I am allowed to say this, being one of them), peace and harmony seldom reign in academe. During the 1990s a particularly nasty civil war raged among American university faculty. As with all academic conflicts, the issues seemed arcane to outsiders. For insiders, the debate concerned the very mission of higher education, and, even more crucially, the basic values of our intellectual culture.

The academic "culture wars" aligned traditionalists against a coalition of postmodernists, Afrocentrists, feminist theorists, and other "tenured radicals" (Kimball 1990). One side argued that the university should pursue its traditional role as the guardian of objective knowledge and disinterested inquiry. The other side argued that "objective knowledge" is an oppressive illusion, that *all* knowledge is inevitably political, and that universities should therefore structure the curriculum to empower the marginalized. The "culture wars" have encompassed the "science wars," the equally rancorous squabbles over the role and nature of science. Here the battle lines can be drawn between two broad groups I call "rationalists" and "constructivists." At bottom, the disagreement between rationalists and constructivists is about how scientific communities achieve consensus.

One remarkable feature of science is that views regarded as extremely implausible, even heretical, by a scientific community at one time can become almost universally accepted within a few years. For instance, the concept of continental drift was rejected by nearly all highly regarded geologists during the first half of the twentieth century. Then, in the 1960s, it was rapidly assimilated and quickly became the consensus view in the earth sciences (Hallam 1989). Similarly, *The Origin of Species* received a rough reception from many reviewers

## Are Dinosaurs Social Constructs?

when it appeared in 1859, yet well before Darwin's death in 1882, evolution was accepted as fact by most scientists (see Hull 1973 on the scientific reception of evolution).

Rationalists think that physical reality ultimately drives consensus. They might admit that theories start off, as Einstein said, as free creations of the human intellect. Also, they do not deny social, political, and other external influences on science; instances like the Lysenko affair in the Soviet Union show how ideology (especially when backed by terror) can force consensus. But rationalists hold that in the long run (and sometimes it is a rather long and circuitous run) science can transcend ideology and politics and achieve the rigorous constraint of theory by careful observation of or interaction with the natural world.[1]

It follows that rationalists affirm the existence of an external, independent (of human wishes or concepts), non-socially constructed, physical world which is at least partially knowable; that is, we can observe (either directly or with instruments), measure, and experiment with that world and thereby ascertain certain facts about it. These facts sufficiently constrain our theorizing to warrant our confidence that some of our theories are approximately accurate—or at least empirically adequate—representations of the world.

Constructivists radically oppose this rationalist image of science. They insist that science, like everything else, is governed by rhetoric, ideology, politics, vested interests, and other social factors. More deeply, they reject the whole epistemological basis of rationalism. They see the claims of science not as "facts" or "discoveries" but as "constructs" created in accordance with the historically contingent linguistic and social practices of scientific communities. Science is not guided by objective methods, but follows arbitrary "rules of the game," conventional practices that have accrued by custom and historical accident. Theory cannot be objectively constrained by fact if the "facts" themselves are merely cultural posits—conventional stipulations of scientific communities. Matt Cartmill uses provocative language but gives an essentially accurate statement of the constructivist view of science:

> The philosophy of social constructivism claims that the "nature" that scientists pretend to study is a fiction cooked up by the scientists themselves—that, as Bruno Latour puts it, natural objects are the *consequence* of scientific work rather than its *cause*. In this view,

> the ultimate purpose of scientists' theories and experiments is not to understand or control an imagined "nature," but to provide objective-sounding justifications for exerting power over other people. As social constructivists see it, science is an imposing but hollow Trojan horse that conceals some rather nasty storm troopers in its belly. (Cartmill 1999, p. 43; emphasis in original)

More formally, I see constructivists as committed to one or both of the following theses:

> *Relativism Thesis (RT): All epistemic standards, including those of natural science, are necessarily relative and parochial. All such standards reflect only the epistemic conventions of particular social groups. No set of such conventions is objectively better than any other.*

> *Nonrationality Thesis (NT): Even when "rational" and "objective" standards are in principle available, scientific consensus is not reached on the basis of such standards. Rather, consensus is a product of conflict and negotiation in which rhetoric, politics, and other "nonrational" social factors determine the outcome.*

A corollary of both theses is that any putative nonsocial physical reality has negligible bearing on the formation of our beliefs.

These two claims RT and NT are not always clearly distinguished in constructivist writings; some emphasize one over the other. I shall call "skeptical" constructivism the view expressed in the RT that it is in principle impossible that there could be objective (i.e., universally valid) epistemic standards underlying scientific belief. I call "cynical" constructivism the view expressed in the NT that it is irrelevant whether objective standards exist since scientific controversies are never (or hardly ever) settled by such standards. Rationalists oppose both the "skeptical" and the "cynical" strains of constructivism: That is, rationalists deny *both* the claims asserted in the RT and NT.

Historian Jan Golinski denies that constructivism should be identified with extreme relativism and the consequent severing of science from material reality:

> The term [constructivism] draws attention to the central notion that scientific knowledge is a human creation, made with available material and cultural resources, rather than simply the revelation of a natural order that is pre-given and independent of human action. It should *not* be taken to imply the claim that science can be entirely reduced to the social or linguistic level, still less that it is a kind of

collective delusion with no relation to material reality. "Constructivism," as I shall characterize it, is more like a methodological orientation than a set of philosophical principles; it directs attention systematically to the role of human beings, as social actors, in the making of scientific knowledge. (Golinski 1998, p. 6; emphasis in original)

However, some social constructivists *do* explicitly espouse extreme relativism and skepticism. See, for instance, Steven Woolgar's remarkably truculent little tome *Science: The Very Idea* (1988) or some of the claims in Harry Collins's *Changing Order* (1985). Those not explicitly relativist, but who, in Golinski's words, direct "attention systematically to the role of human beings, as social actors, in the making of scientific knowledge," are tantamount to what I call "cynical" constructivists. Obedient to their methodological commitments, they consistently deny, dismiss, or ignore the role of epistemic criteria, the traditionally *rational* factors, in the "making" of scientific knowledge.

The war between rationalists and constructivists has been protracted and bitter. Constructivists have recently suffered a humiliating defeat. In 1996 the journal *Social Text,* a leading medium of postmodernist ideas, published the essay "Transgressing the Boundaries" by physicist Alan Sokal. The essay, apparently an argument for constructivism, was actually a hoax, intentionally loaded with gibberish and scientific howlers that any competent reviewer should have caught. Sokal rightly concluded that his successful hoax revealed the scientific ignorance and general intellectual insouciance of the postmodernist science critics (Sokal and Bricmont 1998).

Some have speculated that acrimony surrounding the Sokal hoax may have cost Princeton historian of science M. Norton Wise an appointment at the prestigious Institute for Advanced Study (McMillen 1997). If true, Professor Wise was not the first to get a cold shoulder from the institute because of constructivist associations. Six years earlier, Bruno Latour had withdrawn his application after scientists and mathematicians at the institute had raised objections to his candidacy.

Mathematicians and scientists are not the only ones with strong opinions about Latour. Philosopher of science Philip Kitcher has said about him:

There's a marvelous remark that Russell makes in his *History of Western Philosophy* (1946), when he compares Nietzsche to the mad Lear who says "I shall do such things / I know not what they are, but they shall be / The terror of the earth." I often get the same feeling when I read Latour. It's wonderfully entertaining, but . . . There is no real argument in Latour, just rhetorical flourish and, I'm afraid, serious misunderstandings of lots of views. (Kitcher, quoted in Callebaut 1993, p. 219)

One of Latour's rhetorical flourishes, comparing his role in science studies to Darwin's in biology, elicited the following response from philosopher of biology Michael Ruse: "Frankly, I am well nigh speechless before one who . . . thinks of himself as a 'Darwin of science.' Gross and Levitt [authors of the controversial work *Higher Superstition*] complain of the arrogance of people in science studies. By God, I think they may have a point" (Ruse 1995, p. 9).

Who is Bruno Latour, and why are they saying such terrible things about him? As I mentioned in the introduction, Latour typifies the postmodernist spirit. Irreverent, obscure, intentionally provocative, many of his pronouncements challenge the deepest ideals of science and philosophy (or so it seems—interpreting Latour is not for the faint of heart). *Laboratory Life: The Construction of Scientific Facts,* the book that made him famous (or infamous), was coauthored with Steven Woolgar in 1979. It was here that Latour made many of the statements that have piqued his critics. Though the book is now over twenty years old, it was one of the founding documents of the now burgeoning field of Science and Technology Studies. Its methods and its content remain highly relevant to such studies, so I turn now to an examination of it.

Science often makes factual claims: Quasars are extremely active galactic nuclei; hotspot volcanoes are located over mantle plumes; the most common isotope of carbon has six neutrons; the codons GAA and GAG code for glutamic acid; ammonites went extinct at the end of the Cretaceous. These claims are valued for their own sake as adding to our knowledge of the world. Also, theories and hypotheses are tested against facts. Every scheme of confirmation or corroboration assumes that the rational assessment of theories involves the checking of theories against facts.

Of course, "the facts," and their putative role in the justification of theories, have been under critical scrutiny for some time in the philosophy of science—at least since arguments on the theory-ladenness of perception were set out in Norwood Russel Hanson's *Patterns of Discovery* (1958). Hanson challenged the positivist assumption that observation discloses a theory-neutral realm of facts that permit unbiased arbitration of scientific disputes (for critiques of Hanson and the alleged skeptical consequences of theory-ladenness, see Brown 1989; Scheffler 1982; Kordig 1971).

For Latour and Woolgar, as I document below, facts are not merely theory-laden, they are *constructs* established by *rhetorical* means within the particular cultural milieu of scientific linguistic and social practice. As such, scientific factual claims have no more right to be considered "objective" representations of reality than an alternative "mode of discourse" generated within a contrasting cultural setting. It follows that it is no more or less rational to evaluate theories by testing them against "facts" than by gazing into crystal balls or consulting Ouija boards.

How did Latour and Woolgar arrive at such conclusions? As told in *Laboratory Life,* Latour spent two years observing the activities of scientists in the laboratories of the Salk Institute. His aim was to play the role of the anthropologist watching scientists in their native habitat, like Malinowski observing the Melanisians.

One native ritual Latour found particularly interesting was the bioassay, the process whereby scientists think that they detect elusive substances in tissue. Unlike the scientists, who believe that their bioassays detect substances that are really there, existing independently of the laboratory and its devices, Latour and Woolgar conclude that the bioassay *constructs* the "facts" it allegedly discloses:

> Without a bioassay . . . a substance could not be said to exist. The bioassay is not merely a means of obtaining some independently given entity; the bioassay constitutes the construction of the substance. Similarly, a substance could not be said to exist without fractionating columns . . . since a fraction only exists by virtue of the process of discrimination. . . . It is not simply that phenomena *depend on* certain material instrumentation; rather the phenomena *are thoroughly constituted by* the material setting of the laboratory. The artificial reality, which participants describe in terms of an objective entity, has in fact been constructed by the use of inscription devices. (Latour and Woolgar 1986, p. 54; emphasis in original)

Elsewhere they assert: "We . . . need to stress the importance of not 'reifying' the process by which a substance is constructed. An object can be said to exist *solely* in terms of the difference between two inscriptions. In other words, an object *is simply* a signal distinct from the background of the field and the noise of the instruments" (pp. 126–27; emphasis added).

What are we to make of these passages? By "inscription device" Latour and Woolgar mean any piece of laboratory apparatus that produces symbolic output, such as graphs, charts, or numbers. As J. E. McGuire notes, Latour and Woolgar are presumably not making the trivial claim that there is no instrument output without instruments (McGuire 1992, pp. 169–70). What then is their constructivist claim?

In reading Latour and Woolgar, I discern three nontrivial senses in which they can be taken as asserting that scientists "construct" their "facts":

> *1. Scientific objects are texts, such as the symbolic output of an inscription device or the written records of scientists. There is no objective physical reality that those texts are about. A scientific object just is the text itself. A biochemical substance, for instance, is not detected by an inscription device; it is merely the symbolic output of that device. On this construal, scientists construct reality because there is nothing transcending the texts they produce.*
>
> *2. The physical objects studied by science are more than texts. Moons, microbes, etc., really do exist. However, those objects are somehow created by scientists. For instance, Galileo created the Galilean moons of Jupiter ex nihilo; Pasteur likewise created the anthrax microbe. In other words, Latour and Woolgar are asserting a form of idealism that claims that the physical realities studied by science are constructed in the sense that they are literally brought into being by scientists.*
>
> *3. Science does not discover extralinguistic facts, it generates fact-statements. Nature (conceived as an external, independent, nonsocial reality) in no way constrains the beliefs expressed in such statements. On the contrary, "nature" is whatever scientists agree to regard as natural, and "reality" is a product of convention. Scientific fact-statements are constructs generated by the contingent social and linguistic practice of scientific communities.*

Much of the uncertainty about how to interpret talk about constructing facts arises from the ambiguity of the word "fact." "Fact" can refer either to extralinguistic realities or to fact-statements, statements intended to assert truths about external realities. Senses #1 and 2 are ontological claims—assertions that certain putative objects really

## Are Dinosaurs Social Constructs?

do or do not exist. Sense #1 denies the existence of extralinguistic scientific objects; scientific objects are texts. Sense #2 says that such entities exist and are somehow created by scientists. Whatever its ontological implications, sense #3 makes an epistemological claim, that is, a claim about the grounds of our scientific knowledge. It claims that objective physical entities, if there be any, play no role in the construction of scientific knowledge.

Textual support for each of the above interpretations can be found in the writings of Latour and Woolgar. In particular, they frequently seem to assert sense #1 by claiming that scientific objects *are* simply the output of "inscription devices." However, I think something like sense #3 would be Latour and Woolgar's most plausible (or least wildly implausible) claim, and I shall settle on a modification of this sense as my reading of the main constructivist claim as presented in *Laboratory Life*.

Much of Latour and Woolgar's argument is summarized in the following quotation:

> From their initial inception members of the laboratory are unable to determine whether statements are true or false, objective or subjective, highly likely or quite probable. While the agonistic process [i.e., scientific debate] is raging, modalities are constantly added, dropped, inverted, or modified. Once the statement begins to stabilise, however, an important change takes place. *The statement becomes a split entity.* On the one hand, it is a set of words which represents a statement about an object. On the other hand, it corresponds to an object in itself which takes on a life of its own. It is as if the original statement had projected a virtual image of itself which exists outside of the statement. . . . Previously, scientists were dealing with statements. At the point of stabilisation, however, there appears to be both objects *and* statements about these objects. Before long, more and more reality is attributed to the object and less to the statement *about* the object. Consequently, an inversion take [sic] place: the object becomes the reason why the statement was formulated in the first place. At the onset of stabilisation, the object was the virtual image of the statement; subsequently, the statement becomes the mirror image of the object "out there." (Latour and Woolgar 1986, pp. 176–77; emphasis in original)

Are Latour and Woolgar really saying anything here other than that scientists switch from second order to first order speech as they gain confidence in a hypothesis? First order speech makes statements about things; second order speech is statements about statements. When a hypothesis is first proposed, since no one knows whether it is true or even probable, it is proper to speak of it in a mode that keeps its hypothetical nature firmly in mind. In debating such a hypothesis, scientists are engaged in second order speech, making statements about the statements expressed in the hypothesis. Later, as the hypothesis achieves confirmation, scientists begin to assert the claims of the hypothesis in a first order mode.

Such linguistic practice is a perfectly innocent, ordinary, and everyday procedure. Any hypothesis, even the most mundane, will be treated in this way as we move from doubt to belief. How could it be otherwise? By speaking of a fact-statement becoming a "split entity" or projecting a "virtual image of itself," Latour and Woolgar are using pretentious language to describe a prosaic linguistic practice.

Further, Latour and Woolgar are not simply pointing out, *qua* anthropologists, that scientists fail to notice that they have shifted from second to first order speech. They hint at something much more sinister: "*Our argument is not just that facts are socially constructed. We also wish to show that the process of construction involves the use of certain devices whereby all traces of production are made extremely difficult to detect*" (Latour and Woolgar 1986, p. 176; emphasis in original). This passage insinuates that there is something epistemologically dubious about the shift from talk about hypotheses to talk about facts. Not only are facts "constructed," the "devices" used for their "production" are obscured. To see just what Latour and Woolgar mean, we must follow their account of how facts are "constructed" out of hypotheses.

When a new hypothesis is first proposed, it undergoes a probationary period as its proponents try to convince skeptical scientists. Latour and Woolgar call this the "agonistic" process (p. 237). During this process, which is essentially one of *rhetorical negotiation,* the proponent tries to get other scientists to drop the various modalities, for instance, "possibly," "maybe," "could be," etc., employed to qualify the hypothesis and emphasize its probationary character (on "negotiation" see pp. 155–67 of *Laboratory Life*). The aim is to get other scientists to stop regarding the claim as hypothetical and instead to see it as a literal expression of a reality that is "out there." Latour and Woolgar emphasize that scientists involved in such negotiations are not so much concerned with evidence per se as with what their skeptical interlocutors will *take* as good evidence (p. 156).

They seem to mean something like this: Scientific discourse is regulated by certain rhetorical rules. These rules dictate that proponents of disputed hypotheses must mention the instrument readings, procedures, methods, and so on regarded by scientists as relevant to the hypothesis evaluation process. However, once such persuasion is accomplished, that is, once "stabilisation" is achieved, no more mention is made of these instrument readings, methods, etc.—which Latour and Woolgar regard *merely* as rhetorical devices. From this point on, hypothesis-statements are transformed into fact-statements, and all references to the rhetorical means (i.e., methods, techniques, instrument readings, etc.) whereby this transformation was accomplished are omitted. According to Latour and Woolgar, an "inversion" has occurred in which the actual rhetorical processes whereby scientists are led to believe in a supposedly objective reality are obscured, and it comes to be thought that, instead, it is the putative reality that justifies the statements (p. 176).

For the scientist, such an "inversion" is unproblematic. Once a hypothesis is confirmed to the satisfaction of the relevant scientific community, then continued reference to the confirming evidence would be pointless. Should we continue to mention Harvey's experiments whenever we speak of the circulation of the blood? Nevertheless, Latour and Woolgar imply that scientists are mistaken when they think that their practices of "confirmation" amount to *more* than rhetoric. Scientists are certainly often aware of the rhetorical advantages of citing a particular technique or method in their debates with colleagues (more on this below). However, they also generally believe that such techniques are reliable mediators of external reality. For Latour and Woolgar, *this* is the rub. For them, such techniques are *merely* rhetorical devices. Scientists construct their fact-statements in accordance with the accepted scientific social and linguistic practices and then forget that consensus has been *rhetorically* achieved (the wording in this paragraph is mine, based on my best reading of Latour and Woolgar 1986, pp. 176–77).[2]

For Latour and Woolgar, what scientists and philosophers call the "confirmation of hypotheses" is really an epistemological Lethe in which the constructed nature of hypotheses is forgotten. The propositional attitudes of scientists may change, but hypotheses are born as constructs and so they remain. Scientific methods, techniques, and the like are *nothing but* rhetorical devices that help convince scientists that their socially constructed stories are more than just stories.

Having detailed Latour and Woolgar's account of how scientific facts are allegedly constructed, I believe that I can finally offer an interpretation of their main constructivist thesis:

> *Constructivist Thesis (CT): The standards, methods, techniques, etc., deployed in scientific debates are merely sophisticated* rhetorical devices; *there is no reason to think that such methods, techniques, and the like are reliable means of confirming hypotheses about external reality. Scientific fact-statements are linguistic constructs produced by scientists in accordance with the conventions of scientific social practice. Such constructs take on the appearance of factuality when scientists deceive themselves by obscuring the rhetorical nature of the process whereby they achieve consensus.*

CT implies that science is just one form of discourse among many. From CT it follows that there is no reason to think that scientific fact-statements are any more "objective" than those generated by any other set of cultural and linguistic practices. Science, like astrology or voodoo, will have its own rhetorical practices, and success or failure within such fields can be judged only in terms of the practices in that field. Science is just one among indefinitely many language games, and each such game must be judged by its own rules.

So are dinosaurs social constructs? Is *everything* that we presume to know about dinosaurs a figment of the paleontological mind? CT implies that dinosaurs are constructs, in other words, that fact-statements about dinosaurs are linguistic constructs generated by paleontologists in accordance with their particular, arbitrary social practices. *All* factual claims about dinosaurs must be so considered, not just the more extreme or controversial ones. This means that there can be no reason to think that the claims made by vertebrate paleontologists are any truer than, say, the claims made by young-earth creationist Duane Gish in his book *Dinosaurs by Design* (1992). Gish maintains that prior to the Flood of Noah, humans and dinosaurs flourished side by side. He even has a picture of St. George slaying a *Baryonyx!*

If CT is true, Gish's claims are precisely on the same level with the standard view that dinosaurs died out sixty-five million years before the appearance of *Homo sapiens*. Both claims are equally social constructs derived from the linguistic and social conventions of the groups that endorse them.

Latour never indicates that he regards such consequences as a reductio ad absurdum of his view. On the contrary, his later writings go into considerably more detail about the role of rhetoric and power politics in shaping the course of science. Also, in his 1987 book *Science in Action*, Latour is even more explicit in insisting that nature has nothing to do with the achievement of scientific consensus: "Since the settlement of a controversy is *the cause* of Nature's representation not the consequence, *we can never use the outcome—Nature—to explain how and why a controversy has been settled*" (Latour 1987, p. 99; emphasis in original). In his 1988 book *The Pasteurization of France*, Latour compares the process of science to Tolstoy's view of warfare. Just as generals can plot strategy before a battle, so philosophers and scientists can prescribe rules for "rational" and "objective" procedures. However, just as Tolstoy believed that all strategy was forgotten in the "fog of war," so Latour believes that the scientific decisions are not reached in the prescribed "rational" manner, but through "war and politics" (Latour 1988, p. 4).

Here Latour seems to be defending what I earlier called cynical constructivism, the view that, though rational procedures may be available in principle, they are irrelevant to the actual rough-and-tumble process of science. He argues that scientific success goes with the ability to form networks, associations with powerful allies who can deploy overwhelming resources in a controversy: "There is a point where no matter how pig-headed a dissenter could be, enough is enough. The dissenter would need so much more time, so many more allies and resources to continue the dissent that he has to quit, accepting the ... claim as an established fact" (Latour 1987, p. 78). In a scientific debate, those who can form more powerful networks and deploy more resources can simply outgun their opponents. More ominously, if dissenters are too intransigent, they are threatened with professional dishonor or even branded as crackpots. According to Latour, this was the threat John Ostrom faced when he promoted radical ideas about bird evolution (p. 201). In short, dissenters are cowed into silence.

Is this the whole story? If so, then Paul Gross and Norman Levitt are not being unfair when they accuse Latour of claiming that the winner of a scientific debate is the one who assembles the "biggest and nastiest gang of henchmen" (Gross and Levitt 1994, p. 59). Latour encourages such an analysis in a set of aphorisms he calls "Irreductions" appended to his 1988 book. One of these aphorisms says, "There is only one rule: 'Anything goes'; say anything as long as those being talked to are convinced. ... That will have to do *for you will never do any better*" (Latour 1988, p. 182; emphasis in original). Here Latour seems to be rejecting the normative distinction between "rational" and other forms of persuasion. Threats are better than logic if they are more effective: "We cannot distinguish between those moments when we have might and those when we are right" (p. 183).

Latour might reply that these aphorisms are factual and not normative, and that he is only reporting how science *really does* work. Perhaps he is arguing that every purportedly rational procedure is so universally flouted in real scientific controversies that they are comically irrelevant: Preaching standards of rationality to battling scientists is like giving the Sermon on the Mount in a barroom brawl. Latour could claim to have seen many such brawls and to have observed that it is rhetoric wielded by powerful networks that settles matters.

Latour's violent language, his metaphors of war and conflict, his emphasis on intimidation, and his refusal to recognize a distinction between might and right support Gross and Levitt's judgment that for Latour science is a Hobbesian war of all against all (Gross and Levitt 1994, p. 58). Would-be dissenters cave in out of fear that their careers will be rendered solitary, poor, nasty, brutish, and short.

Latour's view of scientific rhetoric seems to have changed from his opinion in *Laboratory Life*. As we saw, in that earlier book, Latour construed the traditional methods and standards of science as the rhetoric employed in the negotiations leading to consensus. By the time he wrote *Science in Action*, even the rhetorical role of such "rational" and "objective" factors seems to have receded. Consider his account of the tactics employed by scientific authors to obviate or weaken opposition:

> Like a good billiard player, a clever author may calculate shots with three, four or five rebounds. Whatever the tactics, the general strategy is easy to grasp: *do whatever you need* to the former literature to render it as helpful as possible for the claims you are going to make. The rules are simple enough: weaken your enemies, paralyse those you cannot weaken, . . . help your allies if they are attacked, ensure safe communications with those who supply you with indisputable instruments . . . oblige your enemies to fight one another. . . . These are simple rules indeed: the rules of the oldest politics. (Latour 1987, p. 37; emphasis added)

Not only is scientific debate sheer politics for Latour, I think he would invert Clausewitz's dictum and say that politics is warfare carried

## Are Dinosaurs Social Constructs?

out by other means. In scientific warfare, rhetoric is just another weapon, and intimidation is better than rational persuasion if it is more effective.

In *The Pasteurization of France,* Latour dismisses as an article of a naive Enlightenment faith the view that scientific decisions can be made rationally rather than through chaotic power politics:

> We would like science to be free of war and politics. At least, we would like to make decisions other than through compromise, drift, and uncertainty. We would like to feel that somewhere, in addition to the chaotic confusion of power relations, there are rational relations.... Few people still believe in such an Enlightenment, for at least one reason. Within these enlightened clearings we have seen developing the whole arsenal of argumentation, violence, and politics. Instead of diminishing, this arsenal has been vastly enlarged. Wars of science, coming on top of wars of religion are now the rage. (Latour 1988, p. 5)

Latour wants to destroy the last vestiges of the Enlightenment "faith" and show that science is governed by luck, politics, power grabs, threats, etc.[3]

Perhaps, though, Gross and Levitt (and I) have been misled by some of Latour's purple prose and taken in by his intentionally outrageous metaphors. Sergio Sismondo offers an alternative reading (Sismondo 1996). He argues that Latour evinces a "creeping realism" that opposes the strict constructivism of other practitioners of current Science and Technology Studies (p. 121). After all, Latour does assert that nonhuman "actants," such as microorganisms, have a role in scientific controversy (Latour 1987, p. 84). He also specifically criticizes those sociologists who invoke "merely social" relations to explain the outcome of scientific controversies and seeks to distance his "actor-network" theory from such views (Callon and Latour 1992, pp. 352–56). He says that sociologists of science must make "one more turn after the social turn" and regard scientific objects as more or less natural under appropriate circumstances (Latour 1992). This apparently indicates that Latour sees the outcome of scientific controversy as more than a merely rhetorical achievement. He seems willing to concede a role to nonhuman nature in science.

These apparent concessions to realism are given with one hand and taken back with the other. Latour insists that interpreters of scientific controversies, even those long settled, must steadfastly refuse to invoke nature in accounting for the outcome of those controversies. For instance, soon after the discovery of X-rays, French physicist René-Prosper Blondlot claimed to have discovered another sort of radiation—the "N-ray." Physicists now universally agree that Blondlot was wrong and that dissenting physicist Robert W. Wood effectively debunked Blondlot's claims. Yet Latour insists that we cannot interpret this incident as implying that Wood was right and Blondlot wrong:

> It would be easy enough for historians to say that Blondlot failed because there was "nothing really behind his N-rays" to support his claims. This way of analysing the past is called Whig history, that is, a history that crowns the winners, calling them the best and the brightest and which says the losers like Blondlot lost simply *because* they were wrong. . . . Nature herself discriminates between the bad guys and good guys. But is it possible to use this as the reason why in Paris, in London, in the United States, people slowly turned N-rays into an artefact? Of course not, since at that time today's physics obviously could not be used as the touchstone, or more exactly since today's state is, in part, the *consequence* of settling many controversies such as the N-rays! (Latour 1987, p. 100; emphasis in original)

In other words, the opinions of current physicists about N-ray detection are an outcome of the settlement of the N-ray controversy and so cannot be invoked to explain the settlement of that controversy. In contrast, consider the account of science popularizer Isaac Asimov:

> In 1903, a reputable French physicist, Prosper Blondlot, reported the existence of a new type of radiation from metal under strain. He and others published many reports on this radiation, which Blondlot termed "N rays," the N standing for Nancy, the French city in which he held his university appointment. There seems no question but that Blondlot was utterly sincere. Nevertheless, the N rays were an illusion, his reports proved worthless, and his scientific career was blasted. (Asimov 1966, p. 696)

This is apparently the sort of unabashed Whiggery Latour condemns. But it is the truth. The N-rays *were* illusory, as Wood amply demonstrated, and the reports *were* worthless. In the light of Wood's successful debunking, the correct account really does seem to be that Blondlot failed because there was "nothing really behind his N-rays." In other words, the N-rays Blondlot detected were an artifact of his apparatus and procedure, not an objective natural phenomenon. Wood

# Are Dinosaurs Social Constructs?

demonstrated the artifactuality of Blondlot's reported rays to the satisfaction of the physics community, and those reasons remain good for us today.

Historiographic strictures against Whig history are fine and good, but they should not prevent us from telling the truth (much more on this in chapter 6). Of course, we are in a position to *know* these truths only as a consequence of the successful debunking of Blondlot's claims by Wood and others—but so what?

The naive reader, one uninitiated into the mysteries of constructivism, will think that a simple confusion runs through Latour's works. Latour apparently conflates states of affairs with collective judgments about them. The making of a *judgment* is a historical process and is not complete until the jury is in and the verdict rendered. Surely, though, states of affairs themselves are not products of the judging process—unless we assume that all states of affairs are social constructs and that nothing *is* so until we have *agreed* that it is so.

Of course, this is just what Latour assumes. Wood did not *discover* that the reported N-rays were an artifact; as Latour phrases it above, people "*turned* N-rays into an artefact" (my emphasis). This implies that N-rays could just as easily have been turned into "real" phenomena. There simply was no fact of the matter about N-rays until physicists *decided* that Blondlot was wrong, so naturally those "facts" cannot explain that decision.

The upshot is that for Latour nature cannot be an "actant" until the *human* process of consensus production is settled; prior to that, no such actant exists, not even as a *Ding an sich*. If Latour does not assume such constructivism, if he is willing, as Sismondo suggests, to accord a degree of ontological autonomy to non-socially constructed reality, then he will have to concede a point to the Whigs. He will have to concede that—just maybe—Blondlot really was wrong about N-rays, that is, that his claimed observations of N-rays *really were* illusory and artifactual, and not *made so* by a decision of the physics community. Since Latour is unwilling to grant this, his talk about nonhuman "actants" is empty, or at least inconsistent. *Science in Action* speaks of science as having a Janus-face. Better a Janus-face than talking out of both sides of your mouth.

I shall therefore presume that Gross and Levitt got it basically right in attributing a "warfare" view of science to Latour, that for Latour the settlement of scientific controversies is a product of rhetorical combat waged by viciously competitive networks and that the physical world of nonhuman "actants" has a negligible role in fixing our beliefs. How well does this view of science stand up to the evidence?[4] Theories *about* science, like the theories *of* science, are best evaluated vis-à-vis a competitor, so I shall present an alternative analysis. Marcello Pera is also a student of scientific rhetoric. However, Pera does not study rhetoric in the wholly pejorative Latourean sense of rhetoric-as-bludgeon. Pera is a student of Rhetoric, the venerable study pursued by Aristotle and placed along with grammar and logic in the medieval *trivium* of the seven liberal arts. Rhetoric in this sense is the art of finding the best arguments to persuade a given audience. There is no implication that such arguments must be dishonest or manipulative. To avoid confusion, I shall henceforth refer to Rhetoric in Pera's benign sense with an uppercase "r" and rhetoric in the Latourean pejorative sense with a lowercase "r."

As I have interpreted Latour, he would regard any attempt to distinguish between Rhetoric and rhetoric as pointless given the facts of scientific debate. Is this claim right? Could it be that an examination of actual scientific debates will show that, even in the hottest controversy, such a distinction is in fact often respected? That actual scientific debate does make such distinctions is argued by Marcello Pera in his insightful work *The Discourses of Science* (1994).

Pera begins by agreeing with Latour and Woolgar that scientists are not engaged in the disinterested search for capital "T" Truth, but in the practical matter of confuting opponents and convincing colleagues (pp. 46–51). Scientists are Rhetoricians; they look for the arguments that will most effectively persuade, or at least silence, their opponents. Pera documents the great diversity of argument types that appear in scientific discourse (pp. 59–102). These types are so varied because of the many different dialectical circumstances of scientific debate. For instance, if the main obstacle to acceptance of a new theory is its reliance upon a controversial method, the proponent might argue that opponents have already accepted results derived by that method (pp. 81–82). Such a move does not show that the method is correct; it serves the Rhetorical purpose of overcoming an objection.

On another occasion, opponents might complain that a theory is based on conjecture. Pera notes that natural selection was regarded as such a conjecture by some of Darwin's interlocutors (p. 75). Darwin replied by making an analogy between natural selection and the familiar process of artificial selection. The argument by analogy increased the plausibility of natural selection by comparing it to a well-understood concept.

These instances, and numerous others documented by Pera, show that science can be Rhetorical—in the sense that scientists are always looking for the most effective argument—without being pejoratively rhetorical. The most effective arguments need not work by intimidation, fallacies, subterfuge, or ad hominem abuse. On the contrary, Pera emphasizes that Rhetoric is eminently rational in the sense that it proceeds by searching for *shared* premises, data, and frameworks (pp. 142–48). In fact, as James Farlow points out (private communication), scientists who try to win by using such "little-r" rhetorical tactics are pegged as "arm wavers" and regarded with contempt by other scientists.

Pera defends what he calls the "dialectical" view of scientific rationality: "A theory T is rationally acceptable if and only if it is supported by valid arguments, or if the arguments supporting T are stronger than those supporting [competing theory] T'" (p. 144). Pera thinks that his model has several advantages over positivist attempts to prescribe a universal scientific methodology. He also thinks it is *truer* to actual scientific practice than Latour's "anything goes" account:

> This explication has several clear advantages. Compared to methodological rationality, dialectical rationality is ethically more *tolerant* because it is not linked to a single property or to a set of previously established requisites, but rather to a free debate over different properties and requisites. Compared to the rationality of "anything goes," dialectical rationality is more *adequate,* because it does not depend on the whims of authorities or on external social factors. This reflects actual scientific practice, where it may happen that a scientific community prefers T' to T even if T' does not explain more facts, anticipate "novel facts," solve more problems, etc. but where it never happens that it prefers T' if it is not supported by *stronger* arguments than T. (ibid.; emphasis in original)

Of course, the cynic would be happy to argue cases with Pera. For instance, Darwin was not always the high-minded Rhetorician. He employs two quotations, one by William Whewell and the other by Francis Bacon, as epigraphs to *The Origin of Species*. To anyone familiar with the contents of the book, these quotes must appear as rather unsubtle attempts to reassure Darwin's high-church friends that he is simply continuing the tradition of discovering God's works in nature. Though not as nasty as Latourean rhetoric, surely this device is a bit underhanded.

So do the case studies examined in the first three chapters support Latour's warfare model of science over Pera's notion of dialectical rationality? Consider the debate over dinosaur physiology examined in chapter 2. Why did the controversy die down after the AAAS symposium? Was Bakker cowed into silence? On the contrary, Bakker remains a passionate advocate of dinosaur endothermy. Bakker was certainly not intimidated by the paleontological "establishment"; clearly he relishes his role as "heretic." Latour says that after a while enough is enough, and even the most intransigent dissenter will simply be overwhelmed. Bakker is clearly a long way from throwing in the towel.

Perhaps the debate cooled because, as Farlow complains, Bakker stopped publishing on these topics in the professional journals, which made him hors de combat in the eyes of colleagues (Farlow 1990, p. 44). In fact, by 1980 the argument simply seems to have run its course. As Farlow noted in an authoritative 1990 review of the dinosaur energetics controversy, the terms of the debate had hardly changed in the ten years since the publication of *A Cold Look*. Bakker's arguments in his popular 1986 book are mostly elaborations of those in his 1975 *Scientific American* article and his 1980 contribution to *A Cold Look*. Dinosaur posture, anatomy, and locomotion, the biomass ratios argument, and the bone histology evidence remained the main elements of Bakker's case right through his 1986 book.

Bakker would no doubt charge that the vertebrate paleontology "establishment" did indeed close ranks, forming a powerful network of allies opposed to his "heresy." Again, though, there is no evidence that any such network intimidated Bakker or deprived him of the resources to fight back. There is also no evidence that Bakker would not have been heard had he offered new evidence or arguments. In fact, during the 1990s new evidence on dinosaur energetics has been evaluated, some of it perhaps indicating high metabolic rates (Padian 1997b, pp. 555–56). Apparently, then, the debate atrophied because, in the eyes of his colleagues, Bakker had been given a full and fair hearing and had failed to establish his case. Since no powerful new arguments seemed forthcoming, the debate died a natural death; it did not have to be murdered by henchmen of an oppressive network.

Science is actually remarkably tolerant of "heretics," if only because not much can be done about them. Unorthodox scientists cannot be burned at the stake or even removed from tenured positions. Bakker continues to enjoy much more fame and public recognition than do most paleontologists. If the vertebrate paleontology "establishment" has excluded him, he can just thumb his nose and go on his merry way.

Contrary to Latour's caricature, dissident scientists are seldom shrinking violets who wither under the blasts of hostile rhetoric. If anything, people harden their positions in response to rhetorical bullying. Dissenters do sometimes desist from the public defense of their views, even if they continue to harbor private misgivings. Latour needs to show that when this happens, it is not usually because the dissenter perceives that the weight of evidence and argument favors the mainstream view. Also, Latour unreasonably dismisses the possibility that dissenters can be rationally persuaded rather than simply overwhelmed. In short, Latour's analysis grossly undervalues the role of rational persuasion in the settlement of scientific controversies and grossly overemphasizes nonrational motivations as the causes of consensus.

A more realistic view of the way that scientific dissent dwindles is given by Helge Kragh, in his masterful history of the conflict between the big bang and steady-state cosmologies (Kragh 1996). Kragh notes that by 1970 steady-state views had only two defenders left, Fred Hoyle and Jayant Narlikar. However, the small number of dissenters did not mean the end of the controversy:

> The extreme smallness of the steady-state population at that time does not automatically mean that the controversy had ended, for in some cases scientific controversies may go on with only a single scientist challenging the rest of the professional community. What matters is the response of the community, the mainstream scientists. If they feel the challengers' views sufficiently interesting or provocative to engage in a discussion with them (or for whatever other reason), then the controversy may go on. But if there is practically no mutual communication between the two parties—if it is restricted to one-way criticism with no responses—it is no longer reasonable to speak of the disagreement as a controversy. (p. 374)

In other words, dissenters are not silenced, sometimes they just are no longer heard. It is not a matter of whether the mainstream view has a bigger, stronger "network" in its favor. As Kragh notes, controversy can continue even if only one scientist opposes everyone else (Einstein's ever-lonelier opposition to quantum theory comes to mind). What really seems to matter is that the dissenting view is sufficiently plausible or provocative enough to engage the interest of other professionals. On the other hand, genuine crackpots, like the creationists, get no hearing (not from scientists; they go get their own audiences) even though they are often backed by their own well-funded "research" institutions and think tanks and enjoy networks of allies many scientists would envy. So being backed by powerful cadres is neither necessary nor sufficient to participate in scientific controversy.

The incidents recounted in chapter 1 seem to provide better support for Latour's view. Recall that Holland declined to accept Osborn's dare to mount the *Diplodocus*-like skull on *Apatosaurus*. Osborn supported Marsh's original mounting of the *Camarasaurus* skull, and, as head of the American Museum of Natural History, he could command the support of the most powerful network in vertebrate paleontology. As I noted there, though, Holland was bold, even pugnacious, in defense of his views. He never concealed or dissembled his views on the issue, but freely published them and presented his controversial "Heads and Tails" paper at the Society for Vertebrate Paleontology. It seems most unlikely that Holland was deterred from mounting the *Diplodocus*-like head from fear of Osborn's wrath.

In general, though, does not the "wrongheaded dinosaur" story support a constructivist view of science? For forty-five years the Carnegie Museum, and the world, believed in a composite creature, two dinosaurs in one, like the fabulous amalgamations of Greek mythology. If this could happen at one of the world's great natural history museums, and even be accepted by the whole paleontological community, perhaps such monstrosities continue to abound. Everyone now admits that the *Apatosaurus* accepted for forty-five years was a construct, a figment of the scientific imagination. What guarantee do we have that the world's museums are not presently full of such chimeras? (Some paleontologists believe that the controversial *Protoavis*, the alleged Triassic bird, is precisely such an amalgamation.)

It is clear that science does not always proceed in the traditionally rational manner. Negotiation, rhetoric, and power politics certainly

## Are Dinosaurs Social Constructs?

play a large role, larger than most rationalists might want to admit. Nevertheless, reason and evidence, the traditionally "scientific" factors, also molded and shaped every step of the "wrongheaded dinosaur" story. The 1934 decision to mount the *Camarasaurus* head may have been motivated by a desire to impress the public and end the embarrassment of a headless *Apatosaurus*. However, at the time the weight of scientific opinion, including that of leading authorities like C. W. Gilmore, did not oppose such a mounting. In the end, the evidence did win out, just as rationalists expect. Berman and McIntosh's meticulous argument finally placed the issue beyond controversy.

The lesson to draw about science is that science is a very complex and multifaceted process, a process not reducible to *any* stereotype. Like all human endeavors, science is subject to social influences at every level. However, to a greater degree than the vast majority of human enterprises, science incorporates methods and standards that permit the objective constraint of hypotheses via interaction with external, nonsocial reality. Such interaction greatly limits the effects of bias and sometimes elevates scientific hypotheses to a very high level of credibility. The blood *does* circulate, whatever social influences operated on William Harvey.

The last chapter contained an in-depth analysis of the nature of David Raup's "conversion" to the impact hypothesis. There it was shown that there is no evidence that Raup was cajoled, inveigled, or intimidated into agreement. Neither was Raup overwhelmed by a coalition of more powerful colleagues. On the contrary, Raup had long before cemented his reputation as one of the foremost paleontologists; it was the proponents of the impact theory who were the outsiders and the upstarts.

The upshot is that none of the case studies examined thus far supports the constructivist "warfare" model of science Latour develops in *Science in Action* and *The Pasteurization of France*. Though much bitter and sometimes vituperative language was employed in these debates, there is no evidence that the principal actors were ever rhetorically overawed. Neither is there evidence that dissenters were bullied into agreement by powerful opposing networks or that they caved in from fear of professional disgrace.

On the contrary, the disposition of the various controversies examined in the first three chapters tracked rather closely the arguments and evidence available at the time. When the evidence was overwhelming, as when Berman and McIntosh established the diplodocid associations of the *Apatosaurus* skull, the issue was definitively settled. When, after thorough discussion, the evidence was judged insufficient to support Bakker's warm-blooded dinosaur claim, the controversy abated—but was not considered decisively closed. Temporary closure of an inconclusive debate, with openness to new evidence should it arise, is a preeminently *rational* procedure.

To support his "warfare" model, Latour would have to show that, on numerous occasions, the disposition of scientific controversies went *contrary* to the conclusions indicated by the best arguments and evidence available at that time. In any particular instance, he must show that a scientific community's decision about a hypothesis was independent of rational considerations and correlated instead with political exigencies or social agendas. Politics, personal vendettas, and abusive rhetoric certainly were rife in some of the case studies I have examined. However, it seems to me that any unbiased examination of these cases reveals that the arguments and evidence *really mattered* in the disposition of these controversies. At the very least, rationalists can show that in those instances abundant argument and evidence were *there* and that the participants did not *have* to be swayed by intimidation, abuse, etc.

So, in general, does science operate closer to Pera's or Latour's account? A definitive answer would require a historical investigation of a scope vastly greater than any that Pera, Latour, or anyone else could achieve. Innumerable episodes and anecdotes could be adduced for either side. Further, any such judgment can only be qualitative; I can think of no way to quantify the correlation between debate outcomes and the cogency of available argument and evidence.[5]

When the evidence is very complex and/or incomplete, as it is bound to be with respect to any attempt to give a comprehensive theory of scientific rationality, Rhetorical considerations become paramount. In particular, we have to ask on whom the burden of proof should primarily fall. It seems to me that the burden of proof here is definitely on Latour. He is the one apparently making the outrageous claim that scientists are actually a kind of Mafia who extort consensus through rhetorical manipulation and power-politics intimidation. The thesis that there is a moral and rational equivalence between Louis Pasteur and Vito Corleone surely is going to require massive justification.

## Are Dinosaurs Social Constructs?

The case studies examined here are sufficient to show that Latour's warfare model cannot be the whole story. To reject Latourean cynicism it is sufficient to show that scientists *can* and *sometimes do* make rational decisions based on the best available arguments and evidence. Even if scientists are not always angels of dialectical rationality, neither are they always apes of network nastiness.

Latour might reply that Pera and I beg the question against him by assuming that arguments can be evaluated independently of their *actual acceptance*, whereas he has insisted that "anything goes," so long as it persuades. If, in principle, all we can say of arguments is that whatever works works, we cannot distinguish between rational and irrational ways of resolving controversies. "Rationality" is defined by the winners.

Let C be the rationalist claim that some arguments have intrinsic logical or evidential cogency. C seems to be the claim that Latour wants to deny since he wants to reduce cogency to efficacy (i.e., the only possible test of cogency is whether an argument actually persuades). Obviously, Latour cannot *argue* for not-C in the sense of claiming objective warrant for not-C; that would be blatantly self-defeating.[6] He can coherently claim that an argument for not-C is cogent only if it actually persuades, and since it doesn't, it isn't. Put simply, if no argument has intrinsic cogency, then neither do Latour's. There is no escaping this simple logical point. A skepticism that equates the value of arguments with the fact of acceptance therefore undermines itself.

Such reflexive or self-referential paradoxes inevitably arise when global skepticism seeks to undermine our basic logical or empirical reasoning (Nagel 1997).[7] This is why I have *not* interpreted Latour as making the claim that the rational resolution of scientific controversies is impossible in principle—only that *in fact* they never (or hardly ever) are so resolved. So construed, Latour's claim is an empirical model, to be evaluated vis-à-vis the available evidence like any other such model. I have argued that, given the evidence provided in the case studies, Latour's "warfare" model is not well confirmed.

In this chapter we have seen that constructivists like Latour make big, bold claims; they see themselves as offering a sweeping, revolutionary new vision of science. But constructivism is a paper tiger. Earlier I quoted Philip Kitcher's comparison of Latour to the mad Lear. Continuing in the Shakespearean vein, I would *not* say that constructivism is "a tale / Told by an idiot, full of sound and fury / Signifying nothing"—but it does not signify very much. Constructivists do say some true things about science: Science is often shaped by social interests, ideology, and other such factors. Theory is underdetermined by evidence. Experimental results are not always straightforward, but often require extensive interpretation. But these things are old news to philosophers of science. Where it is true, constructivism is hackneyed; where it is original, it is vacuous.

What, then, of dinosaurs? Can we ever be confident that we have "drawn out leviathan," that our theories are true representations of real dinosaurs, and not imaginary constructs invented to serve social interests or ideological imperatives? Even if constructivism is rejected, it would be a very serious error to underestimate the influence of social factors, ideology, and other "external" influences on the course of science. In my first and third chapters I showed in detail precisely how such factors might have influenced theories about dinosaurs. However, the point of this work is that it would be an equally serious error to discount the fact that accumulating empirical evidence, and reasons of the traditional epistemic sort, *really do* matter for the outcome of scientific controversies.

David Young, in his excellent book *The Discovery of Evolution*, strikes just the right note of balance in our interpretation of science; his words can serve as a coda for this chapter:

> The picture of the scientist as an objective spectator has died a natural death, thanks to the work of historians and philosophers of science. It is now clear that even simple observations are not imbibed passively from the external world but are made by a human mind already laden with ideas. The shaping of these ideas is a human activity carried out in a particular social context, with all the frailties and limitations that that implies. This has led some people to the other extreme, in which scientific knowledge is viewed as no more than the expression of a particular social group. On this view there are no such things as discoveries in science, only changes in fashion about how we choose to view the world. However, such a view cannot account for the fact that scientific understanding does not merely change but is progressive. . . . A sensible view of scientific theory must lie somewhere between these two extremes and embody elements of both. Certainly, scientific discovery does not involve a one-way flow of information from nature to a passive, open

## Are Dinosaurs Social Constructs?

> mind. It involves a creative interaction of mind and nature, in which scientists seek to construct an adequate picture from what they see of the world. (Young 1992, pp. 219–20)

Young also makes an analogy between the progress of science and those tests of visual perception in which images are reduced to tiny dots, and random selections of those dots are cumulatively displayed to a viewer (p. 220). At first, when only a few dots have been shown, there are many possible interpretations of the image. However, as the dots accumulate, the range of feasible interpretations narrows until only one is left.

Young's interpretation is not profound. Compared to Latour's Heraclitean utterances, it is not very exciting either. Those committed to the slogan "It is more important for a theory to be interesting than true" will likely prefer constructivism. However, Young's words have the ring of truth, and for me truth is a sufficient commendation for any view.

# 5

# *LE DINOSAURE POSTMODERNE*

Marilyn Monroe, Elvis Presley, and *T. rex:* Their images are icons of our popular culture. Everyone is familiar with scenes of Elvis gyrating or Marilyn with skirt billowing. Depictions of *T. rex* are similarly ubiquitous. A half-dozen images of dinosaurs can be encountered in a casual stroll through a drug store; a shopping mall can be a veritable *Jurassic Park. T. rex* dominates the dinosaur imagery just as it allegedly dominated the fauna of the Late Cretaceous.

Icons begin as depictions of real people or animals, but soon the image takes on a life of its own. It becomes a pure image, unconstrained by the reality it originally represented. The bloated, drugged, middle-aged Elvis could not efface the iconic image. People overwhelmingly voted to memorialize the familiar youthful Elvis on the postage stamp.

Icons once were rare and precious; now they are mass-produced and disposable like soup cans or cola bottles. We play with our icons, freely vulgarizing, satirizing, or trivializing them. Also, iconic images have innumerable incarnations: Anyone swiveling in a sequined suit, with curled lip and oiled forelock is immediately recognized as an "Elvis." What starts as an image of a particular human being becomes an infinitely realizable archetype.

With such cultural icons, it becomes very hard to discern the reality behind the mythical image. Some decades ago New Testament scholars abandoned the attempt (recently resurgent) to reconstruct an accurate picture of the historical Jesus. They concluded that the overlay of tradition and interpretation, going right back to the earliest sources, precluded any effort to distinguish the historical Jesus from the Christ of faith.

What about dinosaurs? Are dinosaurs buried under so many layers

## Le Dinosaure Postmoderne

of myth that the most assiduous digging will never contact factual bedrock? Is the search for the historical *T. rex* like the quest for the historical Jesus—doomed to uncover myth and interpretation at every level? This intriguing and disturbing possibility is raised in a recent book, W. J. T. Mitchell's *The Last Dinosaur Book* (1998).[1]

Mitchell picks up a theme from Bruno Latour. In the early 1990s Latour claimed to repudiate social constructivism (Latour 1992, 1993). He argued that constructivists and rationalists both err in assuming a dichotomy between the natural and the social and differ merely in the extent to which they invoke one or the other in explaining science. Rationalists think that science is ultimately driven by nature; constructivists say it is society. Latour proposed that scientific objects be regarded as "quasi objects," hybrid entities that can be regarded as more or less natural or more or less social, given the circumstance (Latour 1992, p. 282). Though Mitchell cites Latour only once (1998, p. 164), he draws just such a conclusion about dinosaurs: With dinosaurs a clear demarcation between the scientific object and the cultural icon is not possible; the representation and the reality are fused into a hybrid entity.

That image can no longer be distinguished from reality is a salient theme of postmodernist theorists. In an entertaining primer on postmodernism, Glen Ward discusses the view of Jean Baudrillard, a leading postmodernist pundit:

> We might naturally assume that simulation either duplicates or is emitted by a pre-given real. In this sense we might think that simulation and reality have a necessary attachment to each other. But for Baudrillard, this connection has long since snapped, so that simulation can no longer be taken as either an imitation or distortion of reality, or as a copy of an original. In Baudrillard's dizzying cosmos there is no firm, pure reality left against which we can measure the truth or falsity of a representation, and electronic reproduction has gone so far that the notion of originality is (or ought to be) irrelevant. (Ward 1997, p. 61)

In Baudrillard's own words:

> Abstraction today is no longer that of the map, the double, the mirror or the concept. Simulation is no longer that of a territory, a referential being or a substance. It is the generation by models of a real without origin or reality: a hyperreal. The territory no longer pre-

cedes the map, nor survives it. Henceforth, it is the map that precedes the territory—*precession of simulacra*—it is the map that engenders the territory. (Baudrillard, 1988, p. 166; emphasis in original)

Pure images, images that are not *of* anything, have long been known in art, but the electronic media now mass-produce such images. Since the electronic media are now the (virtual) reality in which we live and move and have our being, our world becomes a cascade of free-floating images:

So what we are presented with in the postmodern mediascape is a matter of circulating surfaces: a zone in which signs can function without having to be plugged in to what we think of a fundamental authentic realm of existence. What we have, says Jean Baudrillard, is a *centreless network of communication* which endlessly reproduces and cannibalises its own constant production of simulations. (Ward 1997, p. 65; emphasis in original)

In presenting the dinosaur as a Latourean quasi object, a hybrid of image and reality, Mitchell is therefore giving us a version of *le dinosaure postmoderne*.

An initial reaction to the postmodernists is that they should turn off their televisions and get outside for some fresh air. Mitchell seems to have read a great deal about dinosaurs, but he never indicates what he has *done*. To really understand what paleontologists do, reading is not enough; you have to get out to the badlands and actually dig up some fossils. Shoveling off a meter-thick layer of overburden in the desert sun gives one a robust sense of reality. It also helps to follow a fossil from the field to the laboratory where preparators exercise the utmost meticulousness in removing specimens from their matrix and preparing them for display or study. When you dig something out of the ground, remove the encrustation of a million centuries, and recognize a jaw or femur, there is a palpable sense of connection to a very real, very deep past.

Postmodernists are certainly right that our media-saturated age has far surpassed all previous eras in the promulgation and trivialization of iconic images. We have black velvet portraits of Elvis and a purple *T. rex* who sings saccharinely to preschoolers. But even amidst the pullulating, ever-mutating media images, surely it is unproblematic to distinguish Barney from a real *T. rex*. What exactly then is the problem? How and to what extent have dinosaur fact and dinosaur fiction become inextricably entwined?

Rigorous argument is hard to find in Mitchell's book; large conclusions often appear as non sequiturs. For instance, after making the uncontroversial observation that our visual images of dinosaurs must be imaginative reconstructions since all we ever *see* of them are fragments of fossilized bones, he offers the unwarranted conclusion that commonsense contrasts between "science" and "fantasy" are misleading (Mitchell 1998, p. 56).

Chiefly, Mitchell makes his case by detailing the complexities of the symbolic role that dinosaurs play in modern culture. He holds that the symbolic role of dinosaurs has become so pervasive and fundamental that dinosaurs should be regarded as our modern totems:

> The dinosaur is the totem animal of modernity. By this I mean, first, that it is a symbolic animal that comes into existence for the first time in the modern era; second, that it epitomizes a modern time sense—both the geological "deep time" of paleontology and the temporal cycles of innovation and obsolescence endemic to modern capitalism; and third, that it functions in a number of rituals that introduce individuals to modern life. (p. 77)

This is quite a mouthful. What does it mean to say that the dinosaur is the modern totem animal? Mitchell says that part of what it means is that dinosaurs can be the "clan sign" for all sorts of social collectives. They can represent vanishing races or imperialistic colonizers, brotherhoods or sisterhoods, monstrous ferocity or gentle giantism (p. 78). Dinosaurs are omnibus symbol-bearers, prefabricated metaphors suitable for every occasion.

Some of the symbolic connections Mitchell offers seem plausible, even obvious, like dinosaurs as modern, scientific substitutes for dragons (pp. 87–92). Other connections seem strained and arbitrary, like the one alleged between dinosaurs and the supposed ills of capitalism. (Why not a symbol of the obsolescence of socialism?) In general, when Mitchell claims that X is symbolic of Y, it is hard to see just what constrains these claims. This is especially so if, as he repeatedly indicates, a symbol can stand for a thing *or* its complete opposite *or* something wholly unrelated. If dinosaurs can stand for anything (innovation *or* obsolescence, gentleness *or* ferocity, etc.) at any time, it is hard to see what standards guide interpretation.[2]

## Drawing Out Leviathan

Of course, a symbol can mean different things in different contexts; I am not denying this obvious truth. What Mitchell seems to mean when he attributes symbolic significance to dinosaurs is that dinosaur images manifest subconscious desires, fears, obsessions, etc., that were rife in the culture of the creators of those images. Interpretation will therefore involve articulating that tacit cultural context. But when is a cigar symbolic, and when is it just a cigar? The problem is that for Mitchell (despite his disavowal, p. 177), a cigar is *never* just a cigar. *Everything* about dinosaur images is simply bursting with symbolic significance; every detail becomes a vast repository of nuance and an occasion for an orgy of free association. (I justify this impression below.)

However, my main purpose is not to dispute the details of dinosaur symbology, or the standards (or lack thereof) guiding Mitchell's interpretations, only the epistemological conclusion he draws: "The truth is that the dinosaur is never really separable from its popular and cultural status" (p. 82). Though such claims are scattered through Mitchell's text, it is hard to identify any solid argument for them. Cataloging the proliferation of dinosaur images and asserting their alleged symbolic affinities fails to show that paleontologists cannot isolate a scientific object effectively unpolluted by popular culture.

Consider a case in point: I quote, pretty much at random, from the chapter "Carnosauria" in *The Dinosauria* (Weishampel, Dodson, and Osmolska 1990):

> The femur . . . is usually the longest element of the hindlimb, but in tyrannosaurids, it is nearly equal in length to the tibia. The femur is robust and distally curved so as to be concave caudally even in large carnosaurs such as *Tyrannosaurus rex*. In cranial view, the shaft is straight. The well-defined head projects medially where it is expanded to give a constricted neck connecting to the proximal shaft. The head is oval or hemispherical; it is never compressed. The greater trochanter is lower than the femoral head and extends dorsally from the shaft lateral to the head. Between the greater trochanter and the head, the proximal surface of the femur is cylindrical. (p. 184)

Prima facie it is hard to find any tincture of popular culture in this passage. Even the literature on controversial topics such as dinosaur physiology or extinction contains many passages where, to all appearances, the influence of popular imagery is effectively nil. What

## Le Dinosaure Postmoderne

Mitchell needs to do is go through an article in, say, *Nature* and show in detail that, despite appearances to the contrary, the discussion fails to specify a scientific object distinct from popular imagery. Of course, for laypersons it is difficult to keep such distinctions in mind, nor do I think that scientists are always successful in that regard. However, this does not mean that in principle or even in practice that it is always impossible to sort dinosaur fact from fiction.

Interestingly, Mitchell himself denies that scientific depictions of dinosaurs must be inaccurate. In fact, he rushes in where even Robert Bakker might fear to tread—claiming that our reconstructions of dinosaurs probably approach in accuracy what a Jurassic eyewitness could have reported. In the next paragraph he contrasts with current dinosaur depictions the early, inaccurate reproductions of the mid-nineteenth century (Mitchell 1998, p. 50). Mitchell apparently thinks that progress has been made in the accuracy of our dinosaur representations.

These major concessions to realism suggest that a weaker construal could be placed on Mitchell's claims. Perhaps he is saying only that with our representations of dinosaurs there is no clear line of demarcation between image and reality, no definite point where we can say that here fact ends and fiction begins. Instead, there is a continuum with fact and fiction pure at the outer ends but intricately mixed in between. Fossil remains described in osteological treatises will occupy one end of the spectrum and Barney the other. The *T. rex* in *Jurassic Park* would be toward the middle. However, this weakened interpretation of Mitchell's claim is hard to square with his repeated assertion that dinosaur fact and fiction are *never* entirely separable.

Mitchell's style is exasperating. The book consists of many short chapters (forty chapters plus two appendixes in 284 pages). Essentially, these chapters are a thematically connected set of vignettes accompanied by impromptu interpretation. Mitchell seems to realize that his potpourri of narrative and commentary might have left his intentions unclear. In his thirteenth chapter—one-third of the way through the book—he pauses to summarize his aims thus far:

> What I have offered so far is a sampling of tales, traces, and theories of the dinosaur. My aim has been simply to awaken the suspicion that dinosaur facts are never entirely separate from fiction, that real fossil bones are inevitable occasions for imaginative projection and

> speculation, and that the true history of the dinosaur as a cultural concept is stranger than any fantasy we might invent. I've suggested, further, that the whole phenomenon of the dinosaur requires an explanation that comes not just from natural science or common sense, but from cultural science—the historical and psychological study of images, representations and narratives and the ritual practices that go with them. I've suggested that the dinosaur can best be understood as the totem animal of modern culture, a creature that unites modern science with mass culture, empirical knowledge with collective fantasy, rational methods with ritual practice. (p. 91)

Many of these claims do not require ninety pages of text to "awaken the suspicion" that they might be so. Surely dinosaur remains have always invited wonder and speculation. Certainly no one will object to practitioners of "cultural science" having a major role in the study of dinosaurs as a "cultural concept." Just why there is a cultural obsession with dinosaur images is a very interesting question and one that deserves elucidation.

Still, just how are we to take the claim that "dinosaur facts are never entirely separate from fiction"? Some pages later, Mitchell offers another elaboration of his position:

> Biology and culture are [not] utterly distinct, separate realms. If we learn anything from the dinosaur, it is that the flow of images, narratives, and concepts goes in both directions. The "life" of the dinosaur is at the intersection of nature and culture, biology and anthropology, the study of genes and of what have been called "memes," those cultural formations (like images themselves) that are *remembered* and passed on in human history. (pp. 106–107)

This appears to assert once again that dinosaurs are natural-cultural hybrids, both scientific objects and cultural icons. To understand these hybrids we need Marx and Freud just as much as we need Darwin:

> But how do we study the life of a cultural icon that is also a scientific object and the assembled relics of an extinct group of animals? What method is appropriate? . . . We need a synthesis of Darwin and those other two great early modern thinkers (now, alas, popularly dismissed as dinosaurs), Marx and Freud. We need Marx to understand the relation of dinosaurs to politics and economics, to the development of capitalism as a world system. We need Freud to grasp the psychological components of dinosaur fascination, its roots in unconscious processes of fantasy, identification, and desire. (p. 107)

Mitchell's phrasing is deeply puzzling. Surely *dinosaurs* had no relation at all to politics and economics; the Mesozoic was blessedly free of those institutions. Further, however useful Freudian explanations might be in understanding *our* obsessions, they seem simply irrelevant to our understanding of dinosaurs. Or should Sophocles have composed *Oedipus T. rex*? In a passage quoted earlier, Mitchell speaks of the dinosaur as "a symbolic animal that comes into existence for the first time in the modern era" (p. 77). He cannot literally mean that the *animal* came into existence in modern times (unless in his lexicon "modern" extends back to the Triassic).[3]

Of course, if we can take these odd ways of speaking as figures of speech, they need not be a problem. Perhaps in speaking of the "symbolic animal that comes into existence for the first time in the modern era" the emphasis is on *symbolic*. Dinosaurs died out at the end of the Cretaceous, but of course their images did not acquire symbolic significance until modern times. So interpreted, the statement is not about *dinosaurs,* but about our modern symbolic representations of them. Instead of being an absurdity, the statement becomes a truism.

However, Mitchell says that the cultural icon *is* the scientific object; he seems to be asserting *identity*. But if the icon and the object are identical, whatever is true of the one must be true of the other, so we cannot affirm something of the symbol and deny it of the creature. To dismiss Mitchell's odd locutions as figures of speech assumes that different things can be true of a thing and its representation, which is impossible if they are identical. But the ability to distinguish between a thing and its representation seems a basic semantic necessity, one that invites total confusion if ignored. When Mitchell speaks of dinosaurs entering into social relations or coming into existence in modern times, he is not just using figures of speech. He is exemplifying the semantic confusion that inevitably results from conflating image and reality.

What Mitchell apparently wants to achieve is the opposite of the process identified by Latour and Woolgar whereby scientists begin to distinguish between an actual entity and its symbolic representation (Latour and Woolgar 1986, p. 176). With dinosaurs, Mitchell wants to reverse this process and thoroughly conflate the representation and the reality. Mitchell speaks modestly of dinosaur facts as never being *entirely* separate from fiction, but how can he say they are separate at all? How does his conflation of image and reality leave us *any* basis for distinguishing dinosaurs from our representations of them? I see no possible answer on Mitchell's terms, and I regard this consequence as a reductio ad absurdum of his whole approach.

Mitchell might concede that an in-principle distinction between objects and their representations is a linguistic necessity, but he could still argue, much more plausibly, that dinosaur *images* are hybrid objects: "The dinosaur image is the intersection of cultural and natural determinants, a crossroads of scientific knowledge, social interests, and psychological desires" (1998, p. 107). He holds that dinosaur images have had a deep influence on scientific as well as popular conceptions of dinosaurs (pp. 55–56). The implication is that if dinosaur images, which are hybrids of fact and fantasy, are crucial for our *scientific* understanding, then paleontological "fact" must have a large admixture of fantasy.[4] Let us take this as Mitchell's main claim.

At first sight, it does not look too plausible. Whole-dinosaur reconstructions of the sort produced by Charles Knight, John Gurche, David Thomas, Rudolf Zallinger, or Stephen Czerkas do not play a very extensive role in the theoretical work of dinosaur paleontologists.[5] Of course, as we saw in chapter 2, popular dinosaur images have inspired many paleontologists, but to conclude from this that such images are crucial to our scientific knowledge is to confuse the contexts of discovery and justification. However, to make a full and fair assessment of the claim that dinosaur images have had a profound effect on our scientific knowledge, we need to follow a portion of Mitchell's extensive narrative of the "evolution" of dinosaur images.

As Mitchell notes, the Victorian dinosaur models, created by Benjamin Waterhouse Hawkins for exhibition (opened June 10, 1854) in the Crystal Palace Park, were fanciful fabrications (p. 126). Hawkins consulted Richard Owen, the leading dinosaur expert of the time, but Owen had only fragmentary evidence. For instance, *Megalosaurus,* an upright, bipedal theropod, was depicted as an elephantine quadruped. Hawkins's aim was to create dinosaur images that would impress and edify the public in ways that Owen's "bare bones" descriptions could not (p. 96). Mitchell further speculates that British national prestige was enhanced by the life-size displays of fossil reptiles indigenous to Britain. Finally, the depiction of dinosaurs as erect, pachyderm-like quadrupeds supported Owen's anti-evolutionary view that reptiles had not progressed over time, but had reached their most advanced state in the distant past (p. 124).

# Le Dinosaure Postmoderne

Owen's four-footed, elephantine *Megalosaurus* (*left*), illustrated by R. Owen in 1854, and Cope's upright, kangaroo-like dinosaur (*right*), from E. H. Colbert, *Men and Dinosaurs,* 1968.

It is certainly easy to show that Hawkins's dinosaur models were hybrids, reflecting the (very limited) paleontological knowledge of the era, but mainly designed to serve various nonscientific ends such as public education and the enhancement of national pride. Certainly Owen was also grinding an ideological ax in sanctioning these images (Desmond 1982; Torrens 1997; more on this in the next chapter). It is not so clear, though, that these images had any significant effect on the practice of paleontology or scientific concepts of dinosaurs.

By the end of 1858, four years after the opening of the Crystal Palace dinosaur display, very substantial portions of a *Hadrosaurus* skeleton had become available in the United States (Desmond 1975, p. 30). Anatomist Joseph Leidy quickly recognized that the animal was bipedal and that Owen had taken much for granted (p. 31). Other clearly bipedal dinosaurs were soon discovered, and by 1866 Edward Drinker Cope was describing them as lithe, leaping kangaroo-like animals (p. 35). However impressive the elephantine Crystal Park dinosaur sculptures were to the public, there is no evidence that they had much impact on the development of scientific conceptions of dinosaurs, or even on the evolution of dinosaur images. On the contrary, new concepts and new images quickly developed, apparently driven mainly by new fossil discoveries (pp. 30–33).

Mitchell notes that T. H. Huxley soon appropriated the bipedal dinosaur as evidence of evolution, emphasizing its avian rather than reptilian affinities (1998, p. 137). At this point Mitchell pauses to reflect on the significance he reads into the taxonomic division (by Harry Govier Seeley in 1887) of Owen's "Dinosauria" into the "Saurischia" (lizard-hips) and the "Ornithischia" (bird-hips):

> "Bird hips" and "lizard hips" became the anatomical skeleton key, not so much for "decoding" or unlocking the dinosaur, but for giving it an equivocal, ambiguous identity. While dinosaurian duality might seem like a disadvantage from the standpoint of popular taxonomy, the sort of animal classifications that laypeople like you and I employ, it actually has the advantage of producing something absolutely novel in the animal kingdom, a new group of vertebrates on the same level as the mammals, reptiles, birds, amphibians, and fishes, and not reducible to any of them. In this sense, the dinosaur was (and is) a definitively *modern* animal, at once a discovery and a creation of modern science, both a "natural kind" and an artificial construction. (pp. 138–39)

Besides noting Mitchell's penchant for drawing sweeping conclusions from scanty argument, what can we say about this passage? Seeley's taxonomic innovation certainly did not produce a "new group of vertebrates on the same level as the mammals, reptiles, birds, amphibians, and fishes." As Kevin Padian points out, all three editions of A. S. Romer's authoritative *Vertebrate Paleontology* (1933, 1945, 1966) list the Saurischia and Ornithischia as orders of Reptilia (Padian 1997a, p. 177): That is, the Saurischia and Ornithischia were not regarded as a new *class* (the level of Aves, Reptilia, Amphibia, and Pisces), but only as *orders* of a recognized class—Reptilia. (Nobody proposed that dinosaurs be regarded as a separate class until Bakker and Galton 1974.) Just how the taxonomic practice of distinguishing the Saurischia from the Ornithischia makes dinosaurs "definitively modern" creatures is not made clear.[6]

We come now to the turn-of-the-century era. The Marxist and Freudian elements of Mitchell's synthetic method are prominent in his analyses of Charles Knight's paintings, completed in 1900 for Chicago's Field Museum, and Henry Fairfield Osborn's displays at the American Museum of Natural History. He construes Knight's famous depiction of combat between *Triceratops* and *Tyrannosaurus rex* as an allegory of rampant capitalism:

> Knight's scenes of single combat between heavily armored leviathans are the paleontological equivalent of that other war of giants,

> the struggles among the "robber barons" in late-nineteenth-century America. This period, so often portrayed as the era of "social Darwinism," economic "survival of the fittest," ruthless competition, and the formation of giant corporate entities headed by gigantic individuals, is aptly summarized by the Darwinian icon of giant reptiles in a fight to the death. The archaic, dragonlike character of the dinosaur combines with its scientific novelty to capture perfectly the paradoxical synthesis of feudalism and modernism in America during this period. (Mitchell 1998, p. 143)

Osborn's displays, on the other hand, allegedly had strong overtones of masculine sexual preoccupations:

> Perhaps Osborn's most important contribution to the myth of the modern dinosaur was his linkage of it to questions of male potency. The connection between big bones and virility had already been established by the Jeffersonian mastodon. Big bones were also trophies of the masculine ritual of the big game hunt, and the phallic overtones of "bones" need no belaboring by me. (p. 150)

Mitchell even reads deep political significance into the greenish color Knight gave his dinosaurs:

> So where does this leave greenness? Is it a symbol of the "colored" racial other, the savage, primitive denizen of the green world? Or is it an emblem of the white man's burden, the color of the military camouflage required for the Great White Hunter to blend in with the jungle and thus to dominate it? The answer by now should be evident: gray-green is both black and white, both "colored" and no color at all. It designates both sides of the "color line." ... The dinosaur can symbolize the dominant "master race" that commands a global empire, the vanished, savage races that lost out in the Darwinian struggle, or an invading horde of aliens who threaten white supremacy. (pp. 147–49)[7]

Frankly, these pronouncements leave me speechless.[8] What any of this has to do with *dinosaurs* is beyond me. In fact, I think these passages tell us less about the political and sexual fixations of Knight and Osborn than of the clichéd preoccupations of contemporary professors of literature. In an attempt to get back to terra firma, let us see what was going on in paleontology at the turn of the century. While, according to Mitchell, Knight was portraying his battling plutocrats and Osborn was displaying his phallic bones, what was going on in *science?*

As noted in chapter 1, the era from 1890 to 1910 was an exciting time of major dinosaur discoveries. Outstanding specimens of gigantic sauropods such as *Diplodocus, Apatosaurus,* and *Brachiosaurus* were collected from the American west and Africa. The liveliest paleontological controversy of the period concerned the stance of the sauropods. W. J. Holland, director of the Carnegie Museum of Natural History, restored *Diplodocus carnegii* in an erect posture with legs extended straight down like an elephant's. Other scholars objected strongly.

In 1908 and 1909, respectively, Oliver P. Hay, an American, and Gustav Tornier, a German, published articles in which they argued that *Diplodocus* did not stand erect, as in the Carnegie Museum's restoration, but instead possessed a lizard-like posture, with legs splayed out sideways from the body (Hay 1908; Tornier 1909). Hay adduced anatomical and evolutionary considerations for his claim. He argued that the inner toes of sauropods would contact the ground more effectively if turned outward rather than facing directly forward as in Holland's restoration (Hay 1908, p. 679). Also, the immediate ancestors of sauropods were inferred to have had a crocodile-like, splayed-out posture. Such a posture would ideally suit a slow, stupid, swamp-dweller such as *Diplodocus,* so how could selective pressure favor the development of an erect posture (p. 680)?

Tornier's arguments were similar, but he placed even more emphasis on the lizard-like nature of *Diplodocus.* He claimed that the neck and shoulder show a typically lizard-like character and that the shoulder is actually quite similar to that of a chameleon (Tornier 1909, p. 195). He followed this claim with detailed anatomical comparisons which allegedly demonstrated the similarity of *Diplodocus* to present-day lizards and crocodiles (pp. 195–201). Tornier also mocked the "tip-toed" digitigrade posture of the forefeet required by the upright stance (p. 201). He suggested that the absurdity of tiptoeing sauropods be avoided by positioning the limbs in the low, splayed-out manner typical of reptiles (p. 202).

Holland's reply is remarkable both for its cogency and asperity (Holland 1910). Particularly infuriated by what he called Tornier's "skeletal monstrosity," he directed most of his fire at the German scholar (p. 264). He bitterly derided Tornier as a "closet naturalist" who, with the help of a pencil and a "brilliantly illuminated imagi-

symptomatic of the unmitigated hostility shown by many scientists and some philosophers toward the social study of science. He denies that such scholars aim to debunk science:

> Scholars who link science to culture are motivated mainly, I think, not by a desire to discredit science, but to understand it in a new way. We want to know science in a way that some scientists may not want to hear about, as a concrete social practice carried on by real people in a world of values, interests, influences, and drives. I think that inquisitive scientists—those who are faithful to the spirit of their calling—will welcome this sort of inquiry. (1998, p. 183)

My aim in this chapter has not been to debunk cultural studies of science in general. My claim is particular, not general: that Mitchell fails to deliver the goods. His murky mix of Freud and Marx (who perhaps deserve to be labeled "dinosaurs") provides little elucidation of dinosaur images, and he fails in his larger aim of showing that popular imagery has had a significant effect on dinosaur science. In fact, he hardly even *attempts* to show this. Remarkably, he offers no case studies purporting to show how a specific tenet of dinosaur paleontology was shaped by popular imagery. But the irony is that he *could* have shown this; in fact, *I* show it in chapter 1 of this book!

In my first chapter I argued that the decision to mount the *Camarasaurus* head on the Carnegie Museum's *Apatosaurus* was, in all likelihood, not a scientific decision, but was determined by the desire to present a powerful, imposing image to the public. Once mounted, paleontologists accepted the amalgamation without question. As M. K. Brett-Surman points out (private communication), lacking information about how far the skull had been found from the body, the two or three sauropod experts in the world did not object. Even Romer's *Osteology of the Reptiles* (1956) depicted the *Camarasaurus* skull as belonging to *Apatosaurus*. My disagreement with Mitchell is not over whether such influences *do* occur; they do. Hence I do not oppose (and in fact practice) the study of cultural influences on science, but I deny that those influences are nearly so pervasive and significant as Mitchell supposes. Episodes of the sort reported in chapter 1 are relatively rare. Also, I reject the epistemological conclusion that the cultural influences on dinosaur science are so profound that we can never clearly distinguish fact from fiction or image from reality.

What about Mitchell's judgment that those involved in the cultural studies of science do not seek to discredit or debunk science? As we saw in the last chapter, some of Latour's and Woolgar's statements belie this claim. For instance, it was shown that Latour aims to destroy the "Enlightenment faith" that science can make decisions rationally rather than through "war and politics" (Latour 1988, p. 5). What about Mitchell? Does he simply seek to understand science in a new way, or does he have an antiscientific animus?

Mitchell reveals his own agenda in a chapter in which he rejects Carl Sagan's suggestion that some of our more aggressive and antisocial behavior may be due to the atavistic promptings of the primitive "reptilian complex" of our brains (Mitchell 1998, pp. 199–201; Sagan 1977). Mitchell harshly rejects Sagan's account as a "fable" (p. 203) and no more credible than creationists' claims about Noah's ark (p. 201). It is clear that his hostility is not so much due to any particular demerits in Sagan's theory as it is to the fear that science might usurp from humanists the study of moral, political, and social issues (p. 203). He dismisses sociobiology as a "pseudoscience" (p. 106) and fears that the secret agenda of sociobiologists and their fellow travelers is to stamp out cultural studies in general (p. 284; for a rebuttal of such attacks on sociobiology, see Ruse 1979).

Clearly, therefore, Mitchell *is* actuated by a fear of science, and these fears are evident in his condemnation of Sagan:

> Sagan, for his part, is engaged in the kind of sociobiological moralizing that got a well-deserved bad name during the era of rampant social Darwinism and seems to be rearing its head once again at the close of the twentieth century. The trick is relatively simple: one identifies a common human trait (aggression, ritualistic behavior, the formation of social hierarchies and bureaucracies) and traces it to its biological "cause" in our reptilian nature. Human culture is reduced to a model of animal behavior, and a wide range of political attitudes (typically centered on racial or sexual stereotypes) are rationalized as "natural." (p. 203)

But Sagan was hardly a hard-core sociobiologist. The book he co-authored with his wife Ann Druyan, *Shadows of Forgotten Ancestors*, makes mild and qualified claims about human evolutionary history and its relevance to human behavior. Perhaps human sociobiology has been appropriated by those with reactionary and even racist agendas. However, the categorical rejection of any possible biological basis of

## Le Dinosaure Postmoderne

human behavior also strikes me as an ideologically driven dogma. Sagan himself was a man of very liberal and humane outlook, and Mitchell's treatment of him amounts to little more than a smear.

More importantly, there is and can be no a priori guarantee that science, particularly evolutionary biology, will not have something significant to say about human nature and culture, and what it says *may* conflict with the dearest left-liberal convictions of academic humanists (they might also support those convictions—but there is no way we can know this a priori). As Daniel Dennett puts it (1995), Darwinism is "universal acid"; its corrosive power can threaten all ideologies, feminism as well as fundamentalism. Further, despite the furious attacks of academic humanists, and left-wing scientists such as Richard Lewontin and Ruth Hubbard, biological research into human nature continues apace (see Wright 1998 for a good popular summary of this research).

Mitchell therefore does seem to fear science, and, given his position and convictions, it is understandable that he does so. One cannot help but feel that fear of possible politically unacceptable *results* of science leads Mitchell to seek a subjective realm forever free from the importunate intrusions of science. For Mitchell, that realm is a Marxist-Freudian reverie about dinosaur images and their symbolic connotations. It is true that science will never intrude into this realm. Conversely, and fortunately for those interested in *understanding* dinosaurs, that realm will never intrude into science.

# 6

# HISTORY, WHIGGERY, AND PROGRESS

COMPARISONS ARE INVIDIOUS, especially when they involve the awarding of superlatives. Though humans love to bestow such honors (Nobel Prize for Chemistry, Best Actor in a Leading Role, Collegiate National Champions, etc.), the criteria for rating anything "best" are often questionable. However, a good case can be made for regarding one fossil as the most important ever discovered: the Berlin specimen of *Archaeopteryx lithographica*. Pat Shipman expresses the awe paleontologists feel toward this specimen: "The importance of the Berlin specimen cannot be overstated. It is more than a stony record of an extinct species. It is an icon—a holy relic of the past that has become a powerful symbol of the evolutionary process itself. It is the First Bird" (Shipman 1998, p. 14).

*Archaeopteryx* looks like the perfect "missing link"—a snapshot of the crucial moment in evolution when birds began to diverge from reptiles. It is a remarkable mosaic of reptilian and avian features (see McGowan 1983, pp. 115–18). Because it is such clear and undeniable evidence for evolution, *Archaeopteryx* has often been the focus of controversy. Some anti-evolutionists have even resorted to the desperate measure of claiming that *Archaeopteryx* is a hoax (see Dingus and Rowe 1997, pp. 121–23 for a brief account).

Crackpots aside, controversy over *Archaeopteryx* continues because paleontologists still disagree over its pedigree. Did birds descend from theropod dinosaurs in the Late Jurassic (the time of *Archaeopteryx*), or did birds evolve from crocodylomorph ancestors far back in the Triassic? The roots of this disagreement grow deep into the history of vertebrate paleontology. It has antecedents in the famous mid-Victorian quarrel between Richard Owen and T. H. Huxley. Huxley, "Darwin's Bulldog," and Owen, a staunch anti-Darwinian, sharply disagreed over the nature of dinosaurs and their connection to birds. Owen held that organisms possessed only a limited number of archetypal body plans (reflecting the intentions of the Creator), and that evolution could never transgress the bounds set by these ideal types. For Huxley, *Archaeopteryx* was clearly intermediate between two major types, birds and reptiles, closely resembling small dinosaurs such as *Compsognathus*.

The story of the Huxley-Owen controversy is told by Adrian Desmond in his 1982 book *Archetypes and Ancestors*. This book is one of the earlier efforts to view the history of science from the standpoint of the sociology of knowledge (p. 114). Desmond's approach is to examine the history of paleontology with the aim of showing how social commitments and ideology influenced science: "So my strategy, broadly speaking, will be to investigate how far abstruse debates over mammal ancestry or dinosaur stance reflected the cultural context and the social commitment of the protagonists, and as a result to determine the extent to which ideological influences penetrated palaeontology to shape it at both the conceptual and factual level" (p. 17). In particular, Desmond disputes the traditional view that Owen was a reactionary ideologue whereas Huxley's science was pure and progressive. He claims that Huxley ground just as big an ax as Owen and that his conclusions about birds and dinosaurs were equally self-serving.

This chapter presents Desmond's account of the Huxley-Owen controversy over birds and dinosaurs as an instance of social constructivist history of science. Historian Jan Golinski has written a sympathetic overview of the influence of social constructivism on the recent practice of historians of science (Golinski 1998). He argues persuasively that social constructivism is best defended not with philosophical polemics, but by showing that it has generated new knowledge and deeper understanding of the history of science (p. xi). I certainly agree that constructivist approaches such as Desmond's have produced new knowledge and deeper insight into the workings of science. However, I shall argue that constructivist history is crucially limited because it cannot account for, or explain away, one of the most important aspects of science—that it progresses.

I also ask why recent historians of science have found constructivism so appealing. One reason is that constructivism avoids the cardinal historiographic sin—the writing of "Whig history." Whig history portrays science as a linear ascent from past error to present truth and classifies historical figures as "progressive" or "obscurantist," de-

Richard Owen (*left*) and Thomas Henry Huxley (*right*).

pending on whether they helped or hindered that ascent. Accepted historiographic methodology now demands historical sensitivity, that is, that the past be interpreted "on its own terms" and not from the standpoint of our allegedly superior enlightenment. In my view, an overly strict interpretation of the ban on Whig history has brought excessive constructivism and relativism into the history of science and denied historians the tools for understanding scientific progress.

Desmond's *Archetypes and Ancestors* focuses on the history of paleontology in Victorian London in the years 1850–75. As I said, he is a revisionist seeking to correct the prevailing Whiggish view that damns Owen as the anti-Darwinian "obscurantist" while lauding Huxley's evolutionism and materialism as enlightened and progressive. Desmond interprets both figures in terms of their ideological agendas and vested social interests.

The conflict between Owen and Huxley came to a head over their interpretation of dinosaurs, especially over their views of the relation of birds to dinosaurs. According to Desmond, Owen was particularly concerned to counter the threat he perceived as coming from the radical, atheistic materialism of Robert Grant. Grant, holder of the Chair of Zoology at London University, was the preeminent British defender of Lamarckism and openly provocative in his radical views (pp. 115–18).

Desmond argues that Owen's reconstruction of dinosaurs (Owen had identified and named the order "Dinosauria" in his paper "Report on British Fossil Reptiles, vol. II," published in April 1842) reflected his aim to rebut Grant's Lamarckism (pp. 119–20). Progression within biological lineages was a cornerstone of Lamarckism; in Lamarckian evolution descendants acquire ever more "advanced" characteristics as they evolve. In contrast, as I mentioned in the last chapter, Owen reconstructed dinosaurs as having many mammalian (and hence "advanced") features that their purported descendants, present-day reptiles, do not have. For instance, his *Megalosaurus* and *Iguanadon* were conceived as upright, elephant-like animals with four-chambered hearts and advanced circulatory systems (p. 119). Owen's reconstructions indicated that the "highest" reptiles had lived in the Mesozoic, leaving only a sprawling, "degraded" swarm of lizards as their progeny. Such "retrogression" contradicts the tenets of Lamarckism.

Desmond's case is persuasive. Owen worked with very scanty remains, and his reconstructions have long been superseded. The "scientific" bases for Owen's reconstructions were unquestionably exiguous, but the theological climate was very warm to Owen's concept of "degradation." Small wonder that power, rank, and prestige flowed to Owen as rewards for his scientific promotion of socially acceptable ends (p. 120). In short, Desmond makes a good case that Owen's dinosaurs were imaginary constructs designed to meet an ideological imperative and serve a social agenda.

According to Desmond, Huxley's view of dinosaurs was just as ideologically motivated and constituted as Owen's. Huxley was an evolutionist and a materialist. He held that birds had descended from dinosaurs and accordingly claimed numerous affinities between the two groups. Huxley maintained that dinosaurs were probably warm-blooded and had a birdlike heart and lungs (p. 127; E. D. Cope and R. S. Lull were also among the earliest to consider possible dinosaur endothermy; see Farlow and Brett-Surman 1997, p. 349). He also anticipated Bakker's taxonomic recommendations by proposing the creation of a new vertebrate "province," the "*Sauropsida*," that would encompass both dinosaurs and birds (Desmond 1982, p. 127; Bakker and Galton 1974).

The traditional story is that the discovery of a nearly complete specimen of *Archaeopteryx* in the Solnhofen limestones of Bavaria in 1861, and the discovery in the same deposits of the small birdlike dinosaur *Compsognatus*, provided Huxley with a test case for evolution (Young 1992, p. 144). However, Desmond claims that Huxley was relatively unimpressed by *Archaeopteryx* (Desmond 1982, p. 124). Huxley was committed to the view that major evolutionary breakthroughs had occurred in "pre-geologic" times, that is, prior to the beginning of the fossil record, so in his view birds already flourished in the Paleozoic (p. 128). A Jurassic creature such as *Archaeopteryx* was therefore far too recent to be the "missing link" between birds and reptiles. Desmond cautions that Huxley's view was hardly the "modern" understanding of bird evolution and was in fact deeply enmeshed in contemporary ideological disputes (p. 130).

According to Desmond, Huxley's stand on the dinosaur-bird connection is best understood in the context of sectarian debates between reductionistic materialists such as Huxley and idealists such as Owen:

> We can better appreciate Huxley's actions by treating them from the standpoint of this "sectarian" split. As part of his campaign to break up obstacles to a materialist evolution between 1867 and 1870, he began dismantling Owen's rhinocerine—and antitransmutatory—dinosaur. In a series of lectures to the Royal Institution and Royal and Geological Societies, he subverted Owen's original intention, throwing the reptile into an evolutionary light and making it the ancestor of birds. This renewed attempt to bridge the classes . . . was timely and necessary: a number of geologists . . . thought that evolution might only apply to species, leaving higher categories like families and orders untouched. On the contrary, Huxley hoped to establish that it could transmute one *class* into another. But first he had to overhaul the dinosaur completely; Owen's model was useless for his purpose, fit only to be disassembled and rebuilt to avian specifications. And this Huxley set about doing with great relish. (pp. 123–24)

In sum, Desmond claims that Huxley's view of dinosaurs was just as much a product of ideology as Owen's.

If Desmond is right, the mid-Victorian dinosaur was a social construct. Desmond specifically draws this conclusion:

> Barry Barnes talks about "ideas as tools which social groups may seek to achieve their purpose. . . . " Not only abstract ideas, but ones embodied in concrete form, as in the case of dinosaur reconstructions, make equally effective "tools." In many ways the dinosaur—the *Victorian* dinosaur—is a singularly appropriate choice for study. Its restoration often entailed free use of the imagination—although,

obviously, imagination is anything but "free." It is socially and culturally constrained in a myriad of unrecognized and unspecified ways, and the creative component in any reconstruction may be the door through which ideological influences shape the beast. (p. 114)

Desmond's argument is powerful. He makes a very good case that Huxley's interpretation of dinosaurs is just as socially constructed as Owen's. Further, his analysis is an exemplary application of the methodology of the sociology of knowledge: A piece of science later regarded as "successful" or "progressive" is explained in the same terms as science long regarded as "obscurantist."[1] No Whiggish popularity contest here! Desmond certainly justifies Golinski's claim that high-quality history of science can be done from a constructivist perspective.

How far can we take such analyses? Can we interpret the *entire* history of paleontology as a sequence of ideologically driven struggles culminating in theories that served the interests of powerful social groups? A drawback of such thoroughgoing constructivism, besides the ones noted in earlier chapters, is that it nullifies any notion of scientific progress. Do we today possess more knowledge of the deep past than did Huxley and Owen? How can we if the *entire* history of paleontology, and not just selected episodes, like Huxley versus Owen or Cope versus Marsh, has been nothing but a sequence of noisy, doctrinaire squabbles?

The Achilles' heel of constructivist history of science is its inability to account for scientific progress. If *all* scientific theories are forged in the smithy of ideological combat, and constituted solely to serve local political and social agendas, it is hard to explain why some ideas prove so fruitful for the *practice* of science—and the achievement of practical results—and others do not. Sociological explanations can elucidate the scientific and popular controversies that surrounded the introduction of inoculation, but they cannot even accommodate the obvious fact that Jenner's and Pasteur's vaccines *worked*. How is it, if scientific results are social constructs, that they have the power to, say, cure diseases or vaporize cities?[2]

But do not the last two paragraphs rather blatantly beg the question? Have not post-Kuhnian developments in the philosophy, history, and sociology of science cast deep doubt over the whole notion of scientific progress? Does science really progress, or is it just one damn paradigm after another? To answer these questions we need a clearer understanding of scientific progress.

Defining "scientific progress" is not easy. As Larry Laudan shows, realists, constructive empiricists, and pragmatists each endorse a different notion of scientific progress (Laudan 1990b). Most accounts of scientific progress focus on comparisons between successive theories in a given domain. When theory $T^1$ is succeeded by theory $T^2$, we can claim that $T^2$ is progressive vis-à-vis $T^1$ because we judge that $T^2$ is a more accurate representation of reality (realism), or is more empirically adequate (constructive empiricism), or is more efficient at puzzle solving (pragmatism).

Advances in scientific theory tell only half the story of scientific progress. The other half involves improvements in scientific *practice,* the practical ability of scientists to subject theories to more stringent, thorough, and reliable tests. Generally, successive scientific theories must accommodate continual improvements in the evidential grounds; for example, more and better data become available, more rigorous methods are developed, techniques are appropriated from other fields, and more sophisticated mathematical and computer tools are employed.[3] I agree with Ian Hacking that too much attention has been paid to scientific theory as a finished product and not enough to the development of experimental and other means of evaluating scientific claims (Hacking 1983).

Scientific progress, the achievement of a deeper understanding of nature over time, involves an elaborate back-and-forth interaction of theory and practice. Science does not automatically progress when new and better data are acquired or more sophisticated methods are developed. As J. G. Lennox puts it, "There have been countless episodes in the history of science where piles of precise, quantitative, experimental data have flowed in and science has made virtually no progress until a theoretical advance allowed us to make sense of it" (personal communication). Conversely, theories later regarded as progressive, like Alfred Wegener's continental drift hypothesis, have languished for years until the means of getting the confirming evidence were developed.

Scientific progress therefore presents two aspects, conceptually distinguishable but intricately connected in practice. Theorizing progresses by producing models that are ever more accurate, empirically

## History, Whiggery, and Progress

adequate, or efficient at puzzle-solving. The *practice* of science also advances as methodological developments permit theories to be tested more thoroughly, reliably, and rigorously. Again, these are two aspects of the same process. Theory $T^2$ explains the data set more adequately than $T^1$ *precisely because* methodological innovations have produced new data that $T^2$ accommodates better than $T^1$.[4]

Constructivists cannot explain scientific progress, so they deny its existence. They have focused on the methods and standards whereby theories are judged. According to the constructivists, such methods and standards are not employed to make objective assessments of theory, but are arbitrarily adopted as historically contingent "rules of the game" (recall Latour and Woolgar's claim that the methods of science were merely sophisticated rhetorical devices). In other words, scientific communities simply stipulate that the game of science is to be played by employment of certain methods or adherence to certain standards.

According to Steven Shapin, a leading constructivist historian, constructivists do *not* contend that scientific beliefs are caused by nonrational social factors *rather* than rational methods. They make the much more radical claim that even the most "objective" methods and "rational" standards are *themselves* socially constructed through and through (Shapin 1992). Or, as David Bloor puts it, "The canons of reason . . . are themselves social conventions, e.g., methodological rules-of-the-game contingently adopted (or rejected) by groups of scientific practitioners" (Bloor 1981, p. 359).

In other words, no method is really any better than another, nor is it more rational to adhere to certain standards; these practices and norms were adopted because of social or political expedience. Progress is always progress by some standard, and if the standards are arbitrary, so will be any claim of progress.

We must therefore examine some of the ways that paleontological practice has developed since Huxley and Owen and ask if it is plausible to regard these developments as mere alterations in the "rules of the game." I shall argue that such a conclusion is completely unwarranted. Further, since the constructivists seem willing to admit that there are many prima facie instances of scientific progress (otherwise their numerous debunking efforts would be pointless), I shall place the entire burden of proof on them: that is, if their debunking efforts fail, I shall assume that science does progress.

The most obvious advance in the practice of paleontology is simply that paleontologists have collected ever greater numbers of fossil specimens. Huxley had one specimen of *Archaeopteryx;* now there are seven. The increase in the number of bird and dinosaur fossils since the mid-1800s has been stupendous: For instance, John Ostrom's discovery of *Deinonychus antirrhopus* in 1964 provided what was apparently the closest link yet between theropod dinosaurs and birds. Recently an article in *Nature* reported the discovery in China of two feathered theropods (Ji et al. 1998). Fossils of Mesozoic birds, once very rare, are now much more common, as paleontologists Lowell Dingus and Timothy Rowe note:

> When we were graduate students, it seemed that *Archaeopteryx* was the sole source of information about early bird evolution. Our textbooks said little about other Mesozoic birds simply because not much was known about them. . . . Recently an explosion of new discoveries illuminated the darkness. In the last fifteen years, new Mesozoic birds have been collected in many parts of the world. South America, Spain, and Asia are producing a wealth of new Cretaceous fossils, including complete skeletons of mature birds, entire nests of eggs, embryonic skeletons, and . . . there are new fossils preserving possible feathers. With every professional meeting, it seems, someone announces a new Mesozoic bird. (Dingus and Rowe 1997, pp. 207–208)

Though the dinosaur-bird controversy is far from settled, current discussions have far more information to draw on than did Owen and Huxley.

Whole subdisciplines have recently developed to supply new lines of evidence for paleontologists. Taphonomy, the detailed study of how organic remains become fossils, was not even proposed as a distinct area of study until 1940 (Shipman 1981, pp. 5–6; some taphonomic principles had been recognized earlier: see Fiorillo 1997). Taphonomy is now a flourishing field that employs experimental and observational data to address paleobiological problems (Fiorillo 1997). Reconstructing living communities from fossil assemblages is a difficult and complex task. Taphonomic studies of the patterns of spatial distribution and orientation of fossils, supported by knowledge of fluvial and other effects on bone distribution, can support inferences about paleoenvironments where earlier paleontologists could only have guessed.

method we shall be following a similar path. For us, methodology will be treated solely as a set of formal statements about how to produce knowledge, and not at all as a determinant of intellectual practice. (pp. 13–14)

The methods and norms invoked by Boyle and the experimentalists to justify the demarcation between science and politics are for Shapin and Schaffer just more conventions that beg for a sociological explanation (p. 342).

Is it reasonable to see the many methodological changes in paleontology since Huxley and Owen as merely historically contingent alterations of the rules of the game? Of course, any particular scientific method will have its limitations. No matter how meticulously applied, no single method ever *compels* consensus. Genuine instances of the *experimentum crucis* are very hard to find; room is almost always left for interpretation and debate (and social influences).

Shapin and Schaffer are also certainly right that we should not unreflectively assume the validity or the applicability of methods. Scientists themselves vigorously debate methods and standards, and there is no algorithm for deciding which rules or norms should prevail in a particular case. All too often scientists do piously invoke rules and norms as post hoc rationalizations. Still, despite these concessions and caveats, it is extremely implausible to say that the adoption of the practices and results of taphonomy, paleoichnology, and cladistics into paleontology was purely conventional and not because these fields *do* provide more and better evidence.

Let us make the Shapin/Schaffer claim (S/Sc) explicit:

*S/Sc: The methodological practices prescribed in science are social constructs. All such practices are adopted by scientific communities because they serve local political and social agendas.*

S/Sc is a *causal* hypothesis; it claims that certain beliefs or practices are always determined in certain ways. In particular, S/Sc implies that scientists never adopt a method just because it is better than the available alternatives; that is, scientists never adopt a new method simply because, for example, it is more rigorous or reliable than the old method and thereby allows for the more stringent evaluation of hypotheses. But this is highly implausible. Surely, *if* some new method is better (more useful, reliable, productive of more or better data, etc.)

than old methods, and *if* scientists see this, it is highly unlikely that they adopt the new method for any other reason than the fact that it is better.

Defenders of S/Sc must therefore argue for one or another of the following:

> *a. No method really is any better than any other.*
>
> *b. Scientists cannot recognize a superior method even if they come across one.*

Shapin and Schaffer must affirm something like (a) or (b); otherwise they will have to admit that scientists might sometimes adopt a method just because it is better and not because it is socially or politically expedient. In other words, if Shapin and Schaffer cannot show (a) or (b), they will have a hard time overcoming the default hypothesis that a particular new method was adopted just because it was better.

How could one argue for (a) or (b)? Of course, Paul Feyerabend famously argues against method and for the conclusion that the only rule that consistently holds in science is "Anything goes!" (1975). He maintains that methodological anarchy is and should be the rule in science. In my view, these grossly hyperbolic claims have been successfully debunked by a number of critics (e.g., Newton-Smith 1981; Laudan 1996; Kordig 1971).

Feyerabend's strategy is to go through the history of science showing how prescribed methodological rules have been successfully flouted by major figures like Galileo. However, as W. H. Newton-Smith notes, anecdotes about the failure of method are insufficient to support Feyerabend's epistemological anarchism (Newton-Smith 1981, pp. 133–34). To justify the rejection of a methodological rule, one would need to show the general unreliability of that rule, that is, that it would fail more often than succeed. However, the only way to show this would be by depending upon other methods that we do deem reliable:

> For how do we know that a particular rule has led us to make unfortunate choices? We have no omniscient God to whisper the answers in our ears. Trapped as we are within the scientific enterprise without such a divine road to knowledge, we have no recourse but to make such judgments on the basis of other principles of com-

parison. Thus any historically based attack on a particular methodological rule of the sort being envisaged will presuppose the viability of other such rules. The best one can do through an historical investigation is to take up a single plank of the ship of methodology while the rest remain, for the moment at least, firmly in place. (p. 134)

Therefore the suggestion that we can have grounds for simultaneously rejecting *all* methods is simply incoherent.[7]

Though his claims of epistemological anarchism are insupportable, Feyerabend did point to a difficult problem for the would-be methodologist: If someone proposes a set of universal rules for science, and it is found that the major figures of science *consistently* (and not just occasionally) flouted those norms, this would be evidence that the proposed rules were incorrect. For many philosophers, it is intuitively unreasonable to think that Newton, Faraday, Maxwell, Pasteur, Darwin, Einstein, et al. might have all along been doing it completely wrong. How then do we justify a particular set of methodological norms given the great diversity of actual scientific practice?

Several philosophers have addressed the task of formulating a historically sensitive methodology. Imre Lakatos proposed that normative recommendations be tested by employing them to make "rational reconstructions" of the major achievements of science (Lakatos 1970). A rational reconstruction is an attempt to retell an episode from the history of science in a way to display it as manifesting methodological precepts, and thus as rational. That set of precepts is best which permits the greatest number of episodes to be reconstructed as rational. Similarly, Larry Laudan at one time recommended that we identify scientific episodes which, according to strong pre-analytic intuitions, are the best exemplars of successful science (Laudan 1977). These cases are then used to evaluate accounts of these episodes written from particular normative viewpoints. Those norms are best which confirm the greatest number of our pre-analytical intuitions.

The recommendations of Lakatos and Laudan have come to grief for various reasons (see Brown 1989, pp. 101–15, for a detailed critique). Chiefly, they depend too heavily on our intuitions about the success or failure of scientific episodes. Every educated person has been exposed to stories, usually highly mythologized, of scientific heroes and villains. It could be that our intuitions are the effect of early indoctrination in Whig history and so are not reliable guides to historical reality.

My approach to methodology has been different. Laudan, Lakatos, and Feyerabend were talking about Scientific Method—Method with a capital "M." They had in mind schemes like Karl Popper's falsificationism that sought to specify a universal, monolithic, ahistorical Method for all scientific inquiry. Popper, and the preceding positivists, were attempting to articulate a model of science at the highest level, one that would account for the broadest features of science and the largest-scale developments in the history of science (McAllister 1996, p. 1). I personally am very skeptical of the existence of any universal Method, and I think that Feyerabend and others (e.g., Chalmers 1990, 1999) have made telling criticisms of such projects.

The basic problem with such alleged Methods is that they focus on *part* of rationality and try to elevate it into the whole. True, theories should be fallible; an invulnerable "theory" could never be replaced or improved and therefore should have no place in science. But the effort to elevate falsifiability into a one-size-fits-all, universal Scientific Method is bound to fail. The problem is that if the falsifiability standard is interpreted too strictly, we will have to regard some of the recognized achievements of science as unscientific. If interpreted too loosely, then clearly unscientific enterprises, like "scientific creationism," will count as legitimate science. Being rational is much harder and more complicated than Scientific Methodologists presume, and the line between the rational and the irrational less clear (though it does not follow that "anything goes").

My concern is not with "big-M" Methods, but with lower-level methods, like the assorted techniques employed by ichnologists in the interpretation of dinosaur trackways or by taphonomists in interpreting a fossil assemblage. Such paleontological methods are certainly not universally applicable, though they are often shared with allied fields such as archaeology and forensic science (see Schwartz 1993 for an account of methods common to such fields). The justification of such lower-level methods does not depend on the success of Popper's or Lakatos's or any other big-M Methodology. For instance, the reliability of the biomechanical studies and living-animal observations supporting Alexander's formula (Alexander 1989, 1991) are not discernibly connected to the fate of Popperian falsificationism. Feyerabend may therefore have shown that there is no Scientific Method, but he has not shown that there are no scientific methods.

Returning to Shapin and Schaffer, how could they argue for assumption (a)—the claim that no method is better than any other? We noted above that global skepticism about method is incoherent. To dispute my claim that progress has occurred in paleontological methodology, Shapin and Schaffer would therefore have to take on those innovations one by one and show that they are no better than the old methods (or no method). For instance, they would have to debunk the application of Alexander's formula for estimating speeds from trackways. Ostensibly, this formula provides much solid, reliable data pertinent to questions about dinosaur locomotion. Presumably, Shapin and Schaffer would argue that Alexander's formula seems reliable to paleontologists because they are ideologically motivated so to perceive it, not because it *is* the most reliable method currently available and is recognized as such.

How would they argue this? Would they go head-to-head with Alexander, challenging the biomechanical researches or disputing the observations on which he based his formula? I think it very unlikely that Shapin and Schaffer would enter such a contest and even less likely that they would prevail if they had the temerity. Even if they succeeded in this particular instance, this would leave a host of other accepted methodological practices for them to debunk.

Shapin and Schaffer would have no option but to examine and reject each proffered method one by one. Again, it would be utterly self-defeating to defend global skepticism about objectivity—to claim that there is no objective knowledge, and so no reliable methods for getting it. *Leviathan and the Air-Pump* purports to adduce historical evidence for its claims, and global skepticism would undermine those claims as much as any other. The inevitably self-defeating nature of attacks on objectivity are detailed by Rescher (1997) and Nagel (1997) (both books are reviewed by Parsons 1999).

The prospects for showing (b), the claim that scientists could not recognize a better method even if they had one, seem no better. The question we need to put to the constructivist is this: Imagine that there is a real, physical, at least partially knowable, non-socially constructed world (see Searle 1995, for an argument that there must be such a world). Imagine further that some means are better than others for getting knowledge of that world. For instance, suppose that double-blind tests really are better means of determining the effectiveness of pharmaceuticals than, say, tossing coins or gazing into crystal balls. In this case, why would scientists, in principle, be incapable of recognizing that one method is better than another? Are they under some sort of curse? How could it be that Pasteur could obtain no reliable methods but constructivist historians can? I find no reasonable answer in the writings of Shapin and Schaffer, Latour and Woolgar, or any other constructivists.

I conclude that the practice of paleontology has advanced greatly since Huxley's and Owen's day, and that constructivists have failed to debunk such developments as the mere alteration of historically contingent rules of the game.[8] Further, since I have placed the entire burden of proof on the constructivists, I shall assume that progress does occur in paleontology. If progress *is* a salient feature of science, then Larry Laudan was justified to complain that recent historians of science have shown far too little interest in the topic:

> If he once grants that certain scientific theories or approaches have proved more successful empirically than their rivals, then the historian who disavows any interest in scientific progress is confessing there are some facts about the past which he has no interest in explaining. Given that no history can be complete, that alone would not be very distressing. But the progressiveness of science appears to everyone *except the professional historian of science* to be the single most salient fact about the diachronic development of science. That, above all else, cries out for historical analysis and explanation. (Laudan 1990a, p. 57; emphasis in original)

Why have historians been so leery about explaining scientific progress? The answer seems to be that any historian who treats progress as anything other than a social construct will be thought guilty of writing "Whig history."

What exactly is "Whig history" and why is it the bête noire of historians of science? According to the *Dictionary of the History of Science,* the British historian Herbert Butterfield (1900–79) gave the term "Whig history" its accepted meaning: "Butterfield defined 'Whig history' more generally as 'the tendency in many historians to write on the side of Protestants and Whigs, to praise revolutions provided that they have been successful, to emphasize certain principles of progress in the past and to produce a story which is the ratification if not the glorification of the present'" (Wilde 1981, p. 445). The problems with such historical analyses are obvious:

> The chief failings of this "Whig history" lie in devoting attention to seemingly modern ideas and movements regardless of their importance in their own time, refusing historical understanding to all opposing tendencies. The conservative thinker, the Tory or the Catholic, is often dismissed as dogmatic, bigotted [sic] or superstitious. Ideas of paramount importance in their own time, but alien to the Whig historian, may even fail to find room in his account. Thus the Whig historian makes the present the absolute judge of past controversies and the sole criterion for the selection of episodes of historical importance. (ibid.)

The temptation to write such Whiggish interpretations has been especially strong in the history of science:

> This is because science has appeared to historians to be particularly progressive. Some historians of science have, therefore, seen the present state of scientific knowledge as an absolute against which earlier attempts to understand Nature could be evaluated.... In its crudest form, "Whig" history of science, like its political counterpart, degenerates into a tale of heroes (those who advanced ideas which accord with the present state of scientific knowledge) and villains. (ibid.)

Helge Kragh, in his *Introduction to the Historiography of Science* (1987), cites a number of ways that Whig interpretations have distorted the history of science. For instance, he notes that medieval alchemists used the terms "sulfur" and "mercury" to name abstract principles, not the elements we recognize today. Whig historians think that the alchemists used the same concepts and dealt with subjects comparable to those of modern chemists (Kragh 1987, pp. 94–95). On such an interpretation, alchemy will seem nonsensical, but not if correctly understood in terms of the concepts and practices of the time. Similarly ahistorical is the attempt to formalize past ideas by translating them into modern mathematical idiom (pp. 95–97). This often results in a distortion of the original concepts. Well-meaning Whig historians also sometimes extend the principle of charity too far and attempt to rationalize embarrassing episodes such as Newton's devotion to alchemy or Kepler's practice of astrology (p. 99). Such rationalization, involving the attempt to read anticipations of currently respectable ideas into past practices, also involves Whiggish ahistoricism.

It is easy to see how Whig history got a bad name. The past is another country, and the Whig historian is like the ethnocentrist who laughs at the customs of the Inuit or Yanomamo. Past scientists should not be judged "good guys" or "bad guys" solely by how well their views anticipated current science. Historical sensitivity requires that we evaluate them in terms of the evidence, standards, and concepts available *to them*. In this case, is it not just pure, undisguised Whiggism to suggest that past scientists may have just recognized (while others obtusely failed to recognize) that some methods *were* better? "Better" according to whose standards?

By *ours*, of course. We hold, and are bound to hold, that some means of acquiring knowledge are better than others. As Nicholas Rescher observes, the only alternative is an indifferentist relativism that affects to maintain a complete egalitarianism with respect to all epistemic standards (Rescher 1997, p. 58). I say "affects to maintain" because such an indifferentist relativism is necessarily insincere. To affect indifference about epistemic standards, to pretend that one's own standards are no better than anybody else's, is tantamount to having no cognitive commitments at all (p. 59). Utter nescience is the only alternative to regarding our own standards as authoritative.

We do, we *must* regard some methods as better than others. Suppose then that by our lights method $M_1$ is a more reliable method than $M_0$: for example, Boyle's experimental method *really was* a more reliable, objective, and useful way of studying pneumatics. Suppose further that we find that in the 1660s scientists abandoned $M_0$ in favor of $M_1$. How do we understand this change in scientific methodology? If we judge that the scientists of the 1660s were in the position to recognize the superiority of $M_1$ over $M_0$, for instance, if the arguments they gave for adopting $M_1$ indicate such an awareness, we draw the obvious conclusion: We judge that the scientists in the 1660s, motivated by the desire to do science better, saw that $M_1$ was superior to $M_0$ and adopted it for that reason.

Is such a judgment Whiggish? Whig history is bad because it is ahistorical. Historical figures are judged by an impossible standard—their decisions are evaluated in terms of concepts, standards, and knowledge inaccessible to them. But what do we do when we find that a scientific community, using the most rational means available in *its* historical context, made a decision we *still* think is right? Do we automatically reject such prima facie instances of scientific advance and grasp for sociological or political factors, determined at all costs to explain away the appearance of progress? Such an approach seems as ahistorical as anything ever perpetrated by Whigs. If it is Whiggish for some philosophers to insist on "rationally reconstructing" every episode in the history of science, the determination to offer "irrational reconstructions" of every advance is equally ahistorical.

It follows that David Bloor's famous symmetry principle (Bloor 1991, p. 7)—that all episodes in the history of science, whether "progressive" or not, are to receive the same sort of (sociological) explanation—is untenable. If, as I have argued, progress *has* occurred in science, we have the right, indeed, the obligation, to explain episodes that *were* progressive *as* progressive. Otherwise, we present a distorted picture of the history of science.

So the shoe is really on the other foot. The a priori proscription against interpreting episodes in the history of science as genuinely progressive, even when they were, is an ahistorical prejudice. Laudan states this point eloquently:

> One can sympathize with Butterfield's concern that a tale of victory, told only by the victors, makes for bad history. But in denying that historians are ever justified in recognizing that certain parts of science are better than others, and in asserting that it is no part of the historian's task to explain the conditions which made them more successful, Butterfield (and those historians of science who follow him) would appear to be abandoning the programme of telling the full story of the past to which they are otherwise so deeply committed. (Laudan 1990a, pp. 56–57)

Whiggish triumphalism distorts history, but it is equally ahistorical to follow an a priori methodological bias that precludes understanding of a salient feature of past science.

Strictures against Whig history have become a sort of fundamentalist dogma in current historiography. This dogma has led to the mistaken view that we are never justified in interpreting episodes of the history of science in terms of our current conceptions of correct scientific methodology. The upshot is that we cannot have any genuine understanding of the nature of scientific progress, since progress is relative to a standard, and the appropriate standard is, ineluctably, *ours*.

An overly strict construal of the ban on Whig history therefore encourages analyses exclusively along constructivist and relativist lines. In addition to the deleterious consequences already noted, a pervasive constructivism in the history of science has at least one other bad effect: It creates an unnatural and unnecessary tension between the history and philosophy of science. Although there have been attempts to create a purely descriptive philosophy of science (Toulmin 1972), these are the great exception, and most philosophy of science remains unabashedly normative. In a passage quoted earlier, Shapin and Schaffer note that they approach issues of truth, objectivity, and proper method differently from "much philosophy of science" (1985, pp. 13–14). For them such concepts have no normative or evaluative function, but are merely additional items requiring sociological analysis.

Any issue of the journal *Philosophy of Science* will show that many philosophers disagree. For instance, many philosophers still think that there are objectively better and worse ways of confirming theories and therefore deny that science is merely a social construct. So long as the historians of science are guided by constructivist principles, they will encounter antagonism from philosophers. Conflicts between individuals are unavoidable, but it seems bad policy for two disciplines that have a common overarching aim, the understanding of science, to be in a position of inevitable conflict over fundamental assumptions.

Speaking personally, I find the study of the social relations of science fascinating. Reading something like *Leviathan and the Air-Pump* therefore evokes a mix of strong emotions: I admire the authors' meticulous historical scholarship and their skill at turning a potentially dry narrative into a good story. However, I am constantly irked by what seems to me a dogmatic commitment to a blinkered methodology. It is small wonder that at the end of their book Shapin and Schaffer state: "As we come to recognize the conventional and artifactual status of our forms of knowing, we put ourselves in a position to realize that it is ourselves and not reality that is responsible for what we know" (p. 344). They put themselves in a position to "realize" that

## History, Whiggery, and Progress

knowledge is a social construct because that is what they were all along determined to discover.

I suspect that my reaction is rather common among philosophers, and such attitudes militate against interdisciplinary cooperation. Well, why should historians' attitudes have to change rather than philosophers'? Is it not arrogant for philosophers to demand that historians surrender constructivism, an approach that, as Golinski pointed out, has had great heuristic value? But I think that constructivist history of science will not wither under philosophical fire, but will collapse from within. It will die because, as I argue above, it is just another form of ahistoricism. Progress is one of the most important aspects of science, and constructivism has no tools to understand it. Constructivism will therefore be abandoned because it fails to deliver the goods.

# 7

# BEYOND THE SCIENCE WARS

So what do we really know about dinosaurs? They were big and fierce and are now extinct. What else? Dinosaurs are not social constructs. They really existed, and we definitely know certain things about them. Though they are icons of popular culture, and their images pervade the media, we can still say that some things about dinosaurs are fact and others fiction. Dinosaur paleontology has progressed enormously over the last century and a half. New methods, tools, and techniques—and the sheer volume of recent discoveries—have brought us much more knowledge about the Mesozoic world and its denizens. So, standing here at the start of a new century and new millennium, can we declare the science wars over and get on with the business of understanding dinosaurs and the rest of the natural world?

Things are not quite that simple. Even when hostilities cease, the rift between C. P. Snow's "two cultures"—the literary and the scientific—will be left wider than ever. The consequences of this sundering of our intellectual culture are hard to predict, but they can hardly be good. One result seems fairly certain: The traditional goal of a liberal arts education, the formation of a whole person (not a brainless aesthete or technocratic barbarian), will be much harder to achieve than it already is.[1]

Can scholars in the humanities or social studies say anything about science that would be useful for the professional scientist? An affirmative answer would be the best hope for ending the science wars. We saw in chapter 5 that W. J. T. Mitchell's attempt to erect a "cultural science" alongside paleontology was a bust. But Mitchell was (I think) trying to say something about *dinosaurs,* and, as his failure so emphatically demonstrates, this should be left to the experts.

Perhaps, though, historians, sociologists, and philosophers can say

ontologists. And I made my mind up then and there that I would devote my life to the dinosaurs. (1986, p. 9)

I find it interesting that Bakker phrases his decision as one to devote his life, not merely to science or to paleontology, but "to the dinosaurs." It is as though he dedicated himself not just to study the dinosaurs but to be their benefactor and advocate.

From childhood on, it seems that *visual images* of dinosaurs—illustrations, movies, museum displays—fired Bakker's imagination and commanded his loyalty. He opens an article in *Earth* magazine with a vivid account of the depiction of active, aggressive brontosaurs in the films *King Kong* (1933) and the silent version (1925) of *The Lost World*. Bakker comments that he has always preferred such a view of *Brontosaurus* and that even in high school he had decided that the traditional account of meek and mild brontosaurs was in error (Bakker 1994, p. 27).

Visual images of dinosaurs have deeply affected Bakker on several occasions, and on those occasions he experienced a strong sense of epiphany. Also, it is fair to say that throughout his professional career Bakker has remained intensely loyal to the vision of dinosaur life disclosed by his inner revelations. It is not merely that such experiences inspired or motivated him; many scientists have been inspired and motivated by such personal experiences. Rather, the *content* of his science has remained doggedly, some would say dogmatically, loyal to those visions. The Great Hall experience told him that dinosaurs were active, aggressive, powerful, and energetic; his scientific arguments have served that vision.

Here I must emphasize that Bakker's susceptibility to vivid visual images, and the deep cognitive significance he places on experiences of personal enlightenment, are not idiosyncratic characteristics. These features are typical manifestations of strong social pressures exerted on Bakker's whole generation. We have already mentioned the countercultural emphasis on the importance of intense personal experience. Further, the baby boom generation, of which Bakker is a member, was the first to experience the enormous influence of television.

As noted in the Introduction, the omnipresence of movies, colorful children's books, and especially television meant that children in the 1950s and early 1960s were conditioned to respond to visual images in a way that no previous generation was. I remember my own excitement at age seven watching Pat Boone and James Mason battle the *Dimetrodon* (not a dinosaur, but close enough) in *Journey to the Center of the Earth*. For some, perhaps including Robert Bakker, the images absorbed in childhood have become a lifelong vision. Bakker remains the nine-year-old kid awestruck by the vision of a fantastic world and its wondrous inhabitants.

Why did the hypothesis of dinosaur endothermy draw so much fire? The influence of Bakker's provocative rhetoric has already been mentioned. Further, the period of campus radicalism was very short-lived; by the mid-1970s the movement was essentially over. With the end of the Vietnam War and the draft, students relaxed back into their accustomed hedonism. The society as a whole was characterized by increasing conservatism from the mid-1970s through the 1980s. By 1978, when the AAAS symposium on dinosaur endothermy was held, the scientific establishment also might have felt the need to reassert its authority. Perhaps Bakker's critics were riding the crest of a much broader wave of reaction against radical dissenters.

For whatever reason, perhaps due in part to the increasingly conservative trend of the times (Ronald Reagan's election was only two years away), paleontologists largely turned a deaf ear to Bakker's replies and remonstrances. (Remember that I am attempting to give a sociological analysis here, not the one Bakker's scientific critics would give.) Bakker, in apparent disgust with professional paleontologists, took his case to the broader public through television appearances and writings in popular books and magazines. This, of course, further alienated the paleontological community.

The popular media have obliged by adopting Bakker's athletic dinosaurs. The reason seems obvious. Bakker's lean, mean dinosaurs fit right in with the national jogging and fitness obsession that began in the 1970s. Since that time the ultrasleek, ultrathin model has been the paragon of beauty. Also, one of the stock media dramatis personae has been the beautiful, but formidable, heroine who can run down, tackle, and subdue the villain, all without smearing her makeup. So, fat rubbery Godzilla—the dinosaur as Sumo wrestler—was out, and lean, lethal *Velociraptor* was in.[5]

Presuming that our sociological account is basically correct, what is the moral of the story? Is the lesson that scientists should discipline themselves to judge strictly by the cold, hard facts and avoid at-

## Beyond the Science Wars

tachment to hunches, visions, or intuitions? No, science needs its visionaries, its heretics, its Robert Bakkers—gadflies who stimulate by irritating. Whether their theories are right or wrong is a secondary consideration; their value lies in their ability to provoke debate and inquiry. Dinosaur behavior and physiology has been a main topic of investigation by recent paleontologists. It is unlikely that these issues would have gotten nearly so much attention without Bakker's stimulus.

Had there been no sixties counterculture, there would have been no Robert Bakker—not the one we know—and the paleontology of the past three decades would have been different, probably poorer. So sociology can tell us much about how social circumstances influence science. However, such analyses only come after the fact: Circa 1967, who could have predicted that countercultural influences, filtered through the idiosyncratic consciousness of a particular Yale undergraduate, would ultimately generate arcane debates on the physiology of fossil animals?

As Desmond and many others have shown, consideration of the ideological, religious, and political contexts of scientific episodes elucidates their historical significance. So the discipline of paleontology cannot be understood without knowing something about its history, including the social context of its development. Insofar as paleontologists wish to understand their discipline, they must learn something about its history and sociology.

Can we go further and say that the sociology of science can offer something to the working paleontologist *qua* paleontologist, something valuable for doing science *now*? It is salutary to be reminded often that we all have axes to grind, and that our motives may be due to internalized social influences. Still, it is hard to see how such reminders can do much to help us choose between currently debated theories. It is a good bet that some of these theories are expressions of wishful thinking or ideological bias. The trick is to tell which ones. Lacking a God's eye view, we simply have no choice but to follow our hunches and intuitions, realizing that these have certainly been shaped by our social milieu, but trusting scientific practice to give nature the final say.

The task of learning about past life therefore will probably not be helped very much by studying the social influences on paleontology. For instance, I may be aware that my visceral opposition to John Horner's claim that *T. rex* was a scavenger and not an active hunter (Horner and Lessem 1993; Horner and Dobb 1997) is motivated by deep-seated wishes and complex, culturally induced prejudices against scavengers. (As Calvin of the *Calvin and Hobbes* comic strip observed, it would be so bogus if *T. rex* only ate things already dead!) Still, the only way to approach the question is to argue it out and hope that objective grounds will eventually emerge to settle the issue.

So we are back to the question that all along has divided rationalists from constructivists: How are scientific controversies settled? What governs the emergence of consensus? If the science wars are not to end with the unconditional surrender of one side or the other (an unlikely prospect), the best solution would be to develop a model of science that, insofar as possible, captures the intuitions of both sides. I would like to end this work on an irenic note by endorsing a model of science which satisfies the rationalist's craving for rationality while accommodating the genuine insights of the sociologists. Such a model has already been developed by Richard Bernstein in his insufficiently admired work *Beyond Objectivism and Relativism* (1983).

Let us begin by considering Harry Collins's *Changing Order* (1985), one of the most vigorous and uncompromising statements of social constructivism. Collins argues, surely correctly, that the process of scientific training is largely a process of enculturation. A scientific community is a subculture with its own folkways, mores, and taboos. A young person seeking admission into such a community must undergo a series of rites of passage which serve to acculturate and condition the novice into accepted ways of thinking and acting. By emphasizing the enculturational aspects of scientific training, Collins intends to show that scientific beliefs largely depend upon prior expectations and socialized perceptions. He fleshes out this claim by focusing on a particular instance of attempted experimental verification, Joseph Weber's effort to confirm the existence of gravity waves (Collins 1985, pp. 80–106).

General relativity predicts that certain gigantic cosmic events, such as the formation of black holes, will emit gravitational radiation in the form of gravity waves. Such radiation would be very weak and difficult to detect. In the 1960s Joseph Weber of the University of Maryland attempted to build devices that would detect gravity waves, and in 1969 he reported their detection (p. 81). Weber's original claim was not accepted, but in the 1970s he modified his devices and reported successful detection. Other laboratories attempted to replicate Weber's findings but reported negative results (p. 83).

Collins's evaluation of the attempts to replicate Weber's results draws the conclusion that there is no criterion for success but success: "Proper working of the apparatus, parts of the apparatus *and the experimenter* are defined by the ability to take part in producing the proper experimental outcome. Other indicators cannot be found" (p. 74; emphasis in original). The only way to tell that someone is, for example, a competent laser-builder is to see whether he or she can build a working laser. Applied to gravity-wave detectors, this means that the only way to tell whether anyone has built a successful gravity-wave detector is to see whether it actually detects gravity waves (pp. 83–84).

In short, we do not know whether we can detect gravity waves until we build a device and get the correct outcome. However, "What the correct outcome is depends upon whether there are gravity waves hitting the Earth in detectable fluxes. To find this out we must build a good gravity wave detector and have a look. But we won't know if we have built a good detector until we have tried it and obtained the correct outcome! But we don't know what the correct outcome is until . . . and so on *ad infinitum*" (p. 84). For Collins, the existence of this "experimenters' regress" means that Weber was judged competent by those who accepted his results and incompetent by those who did not: "Where there is disagreement about what counts as a competently performed experiment, the ensuing debate is coextensive with the debate about what the proper outcome of the experiment is. The closure of debate about the meaning of competence is the 'discovery' or 'non-discovery' of a new phenomenon" (p. 89).

In other words, judgments about experimenter competence, the reliability of the experimental apparatus and procedure, and the success of the experiment rise or fall together. For Collins the only way to break this circle and achieve closure in scientific debates is through employment of negotiating tactics and other such "nonscientific" means (p. 143). Closure is achieved through a process of negotiation among the "core set"—the small set of enemies and allies whose interests and expertise make them the recognized authorities on the disputed issue (pp. 142–43). In the absence of algorithmic directions for the settlement of scientific controversies, such negotiating tactics *must* be employed (p. 143). For Collins, as for Latour and Woolgar, scientific debate disguises the actual nature of these tactics; personal and social interests are "laundered" to look like objective considerations:

> The links to industrial and political interests are sometimes subtle . . . and sometimes, as in the case of, say, the debate over the development of new nuclear power stations, they are obvious. The core set "launders" all these "non-scientific" influences and "non-scientific" debating tactics. It renders them invisible because, when the debate is over, all that is left is the conclusion that one result was replicable and one was not; one set of experiments was competently done by one set of experts while the other—which produce the non-replicable results—was not. The core set "funnels in" social interests, turns them into "non-scientific" negotiating tactics and uses them to manufacture certified knowledge. . . . *The core set gives methodological propriety to social contingency.* (pp. 143–44; emphasis in original)

Although the above passage may seem to belie the claim, Collins says he does not regard scientists as charlatans (Collins and Pinch 1993, p. 145). Scientists are only doing what everybody *must* do; the idea that scientific behavior could be any different is a fantasy (Collins 1985, p. 143).

When I first read the above arguments, I took them to mean that critics of an experiment inevitably argue in a circle since, so Collins seemed to imply, any criticism of the experimenter or the experimental design would beg the question by presupposing the failure of the experiment. Further, Collins seemed to imply that consensus is actually reached by irrational means such as cleverness in negotiating tactics and appeal to vested social interests. A number of other commentators read these passages in much the same way. For instance, Alan Chalmers and James Robert Brown have given some excellent arguments against the existence of an experimenters' regress in the case of Weber's gravity wave research—*where the regress argument is interpreted as a charge of circularity* (Chalmers 1990, pp. 72–79; Brown 1989, pp. 88–92).

Chalmers notes that Weber attempted to strengthen his case by reporting the simultaneous reception of signals by detectors thousands of miles apart. Weber also reported a twenty-four-hour periodicity of detection events which would indicate that detection had occurred when the devices were oriented with respect to a particular stellar

## Beyond the Science Wars

source of gravity radiation. However, the discovery of a computer error undermined Weber's claims about the correlation between separated detectors; these claims were further eroded when it was shown that the alleged detection events occurred four hours apart rather than simultaneously (Chalmers 1990, p. 74).

In other words, Weber's critics did not beg the question against him by *assuming* that his experiment had failed. The skeptics appealed to factors like computer glitches and other basic errors that were judged on the basis of common, widely shared, and independently justifiable standards. Brown states clearly how such background knowledge breaks the circle of the "experimenters' regress":

> We have a large number of background beliefs which will tell us what gravity waves are and how they can, in principle, be detected. Moreover, these various background beliefs will also tell us about the nature of the detecting device, how it works, its degree of sensitivity, the conditions under which it might fail, and so on. This breaks the circle that Collins calls the experimenters' regress. It is not the detection of gravity waves (nor the belief that they do not exist) that uniquely determines whether the detector works; rather it is our well-grounded background beliefs which determine what has or has not been observed. (Brown 1989, p. 90)

In 1994 Collins had an exchange with physicist Allan Franklin on the subject of the experimenter's regress (Collins 1994; Franklin 1994). Franklin reviewed the published documents relating to the gravity-wave controversy and concluded that, though the procedures of Weber's critics were not rule-governed or algorithmic, their consensus was entirely reasonable (1994, p. 471). The judgment that Weber had failed to detect gravity waves was based on epistemic criteria, not a product of rhetorical manipulation or appeal to vested interests (p. 487). He finds no place where Weber's critics had to beg the question or engage in circular argument. I quote part of Collins's reply at length:

> Franklin says that I argue that the decision to reject Weber's findings "could not be made on reasonable or rational grounds" [p. 463]. He says that in my view the experimenters' regress is broken by negotiation with the appropriate scientific community, which does not involve . . . "epistemological criteria or reasoned judgment" [p. 464]. Franklin says my argument " . . . seems to cast doubt on experi-

mental evidence and on its use in science, and therefore on the status of science as knowledge.... Thus experimental evidence cannot provide grounds for scientific knowledge" [p. 471]. He says that I express "*surprise* that the credibility of Weber's results fell so low" [p. 472]. I say none of these things.
The following correctly represents my position: "Collins's view that there were no formal criteria applied to deciding between Weber and his critics may be correct. But, the fact that the procedure was not rule-governed, or algorithmic, does not imply that the decision was unreasonable" (Franklin, p. 471). I am delighted that Franklin agrees that the "procedure was not algorithmic"; this is something I have been arguing for years. But I have never suggested that scientists' actions were unreasonable. (p. 501)

On the next page Collins says, "Franklin says he 'will argue that the regress was broken by reasoned argument' [p. 465]. He is pushing at an open door" (p. 502).

Having first read *Changing Order* I found the above passages simply astonishing. Along with Franklin, Chalmers, Brown, and others, I took Collins to be making radical claims that implied the irrationality of scientific consensus-building. How could we have been so wrong? Was Collins being disingenuous, or were his critics (myself included) being foolish? Maybe neither. What seems to have happened is a two-way misunderstanding. Collins is attacking what he regards as the canonical view of science, that is, that algorithmic prescriptions can exhaustively specify scientific procedure, including the evaluation of experimental results. For Collins, the "enculturational" model of learning is the only alternative to an "algorithmic" one (Collins 1985, p. 57). Later in the book, Collins extends this distinction to cover many other aspects of science. For instance, he says that the algorithmic model holds that complete recipes can be given for the performance of experiments and "all that follows"—presumably, the determination of the bearing of the experimental results on theory choice (p. 159). It is the failure of such a "canonical" model of science, that is, the lack of algorithmic recipes for the replication and evaluation of experiment, which necessitates the resort to "nonscientific" factors in scientific debates (p. 143).

But where and when did such a canonical model reign in the philosophy of science? Let us consider an extended passage from

Ernest Nagel's *The Structure of Science* (1961), a good candidate for a canonical model of science:

> Consider . . . the experimental law that the velocity of sound is greater in less dense gases than in more dense ones. This law obviously assumes that there is a state of aggregation of matter known as "gas" which is to be distinguished from the other states of aggregation such as liquid and solid; that gases have different densities under determinate conditions, so that under specified conditions the ratio of the weight of a gas to its volume remains constant; that the instruments for measuring weights and volumes, distances and times, exhibit certain regularities which can be codified in definite laws, such as laws about mechanical, thermal and optical properties of various kinds of materials; and so on. It is clear, therefore, that the very meanings of the terms occurring in the law (for example, the term "density"), and in consequence the meaning of the law itself, tacitly assume a congeries of other laws. Moreover, additional assumptions become evident when we consider what is done when evidence is adduced in support of the law. For example, in measuring the velocity of sound in a given gas, different numerical values are in general obtained when the measurement is repeated. Accordingly, if a definite numerical value is to be assigned to the velocity, these different numbers must be "averaged" in some fashion, usually in accordance with an assumed law of experimental error. (Nagel 1961, pp. 81–82)

Given this eloquent account of the many assumptions and auxiliary hypotheses involved in experiment, Nagel would have to agree that a very considerable degree of interpretive flexibility often goes with the analysis of experimental results.[6]

Rudolf Carnap, another philosopher often identified with a canonical view of science, discusses the impossibility of testing all of the background assumptions that underlie experiment:

> Practical considerations prevent us, of course, from testing every factor that might be relevant. Thousands of remote possibilities can be tested, and there simply is not time to examine all of them. We must proceed according to common sense and correct our assumptions only if something unexpected happens that forces us to consider relevant a factor we had previously neglected. Will the color of leaves on trees outside a laboratory influence the wave length of light used

in an experiment? Will a piece of apparatus function differently depending on whether its legal owner is in New York or Chicago or on how he feels about the experiment? We obviously do not have time to test such factors. We assume that the mental attitude of the equipment's owner has no physical influence on the experiment, but members of certain tribes may differ. They may believe that the gods will assist the experiment only if the owner of the apparatus wants the experiment made and not if the pretended owner wishes it. Cultural beliefs thus sometimes influence what is considered relevant. (Carnap 1966, p. 45)

Where there is no way to test every assumption, exhaustive algorithmic recipes for experiment, even if possible in principle, are impossible in practice. So for Carnap also, a particular cultural context and many assumptions, if only "common sense" ones, must enter into the assessment of experiment.

The canonical view of experiment therefore looks much less algorithmic and far more flexible than Collins supposes. Maybe scientists and philosophers of science have all along recognized that the evaluation of experimental results is not an automatic procedure, but involves much discussion, debate, and analysis of background assumptions and auxiliary hypotheses. Noted philosopher of biology David Hull, in a review of *The Golem,* which Collins coauthored with Trevor Pinch (1993), concludes that Collins has misunderstood received views of science. For instance, Collins and Pinch tell us that "judging the outcome of a test of a theory is not always straightforward. It is not simply a matter of inspecting theoretical prediction and experimental result as some philosophers believe. Interpretation is always involved" (Collins and Pinch 1993, p. 133). Hull comments: "Time and again they [Collins and Pinch] argue that neither experimentation *alone* nor theory and experiment *alone* can settle scientific controversies (pp. 72, 91, 141, 147). In addition, 'there is no logic of scientific discovery,' nor can science provide 'complete certainty' (p. 142). Unfortunately, these tenets have long been part of traditional philosophy of science" (Hull 1995, p. 487).

Perhaps, then, Collins has unfairly stereotyped the received view of science, which all along has conceded a much greater degree of interpretive latitude and uncertainty than he acknowledges. On the other hand, maybe Collins's critics (including myself) have misconstrued him just as badly. Consider Jan Golinski's interpretation of the experimenter's regress:

> Experimental results can only be assessed by reference to a complex set of contextual factors; they are accepted only if the methods are deemed proper, the apparatus appropriate, the investigators competent, and so on. Verdicts on all these questions stand or fall together, and an experiment cannot be accorded a decisive outcome independent of such complex, situated judgments. What this implies is that, because any subsequent reenactment will always differ in some respects from the original experiment, there is always room to argue that it differs in some relevant respect, which makes it not a fair comparison with the original. Although the individual who performs the reenactment may claim to have replicated the earlier experiment correctly ... the original experimenter can always assert that the second experiment is different in some relevant way and hence not a valid replication. Assessments of whether a result has been replicated are thus always matters of judgment. (Golinski 1998, p. 29)

Some might doubt that it is *always* rational to question the results of attempted replications. Still, Collins's views on experiments, as interpreted by Golinski, do not entail circularity or any other form of irrationality in the debates about claims of replication. I think rationalists can certainly agree that "experimental results can only be assessed by reference to a complex set of contextual factors" and that "assessments of whether a result has been replicated are thus always matters of judgment."

Maybe the main point of the experimenter's regress is merely that with alleged gravity-wave detection, unlike laser making, there is no possible *decisive* test of experimenter competence and concomitant experimental success. If someone questions whether I am a competent laser builder, and concomitantly whether I have succeeded in building a laser, I just switch it on and vaporize some concrete. If I cannot vaporize something or achieve some other such obvious effect, my critics rightly draw their conclusions.

With alleged gravity wave detectors, no such simple demonstration is possible. Neither Weber nor his critics could just go to the alleged detector, switch it on, and show that it does or does not detect gravity waves. Whether the apparatus was or was not a reliable detector of the hypothesized fluxes was precisely the point at issue. In such a case, assessments of competence of the experimenter and reliability of the apparatus must depend on the sort of background beliefs and independently justifiable criteria mentioned by Chalmers, Brown, and Franklin. In his reply to Franklin, Collins says that he does not deny the possibility of replication of an experiment. He continues:

> I actually argue that the TEA laser represents an exemplary case of how replication is managed. It is managed by reference to the correct outcome of the experiment. In the case of gravity waves there was no "correct outcome" until the debate was over. That is why replication, *in the TEA laser sense,* could only be accomplished coextensively with the settling of the debate about the existence of high fluxes of gravity waves. (Collins 1994, p. 503; emphasis added)

In other words, the laser case—vaporizing some concrete—represents an *ideal* of replication, and clearly while the debate over Weber's results was ongoing, gravity-wave detection had not been replicated in *that* sense. Yet Collins seems to have left the door open to less decisive but still powerfully persuasive rational assessment. Collins and his critics would probably still disagree about what should count as "rational" persuasion, but the door is open to further discussion.

What about Collins's claim that to achieve closure "non-scientific" factors *must* be employed (Collins 1985, p. 143)? The disagreement here may be more apparent than real, turning on a semantic point about what counts as non-scientific. If by non-scientific Collins means that scientists must be persuaded by appeal to vested social interests, then there are genuine grounds for disagreement. In fact, I have argued in earlier chapters that epistemic factors *can* be sufficient and that scientists do not *have* to be bribed with the tacit or explicit appeal to vested interests. On the other hand, if for Collins non-scientific is co-extensive with "non-algorithmic," then we can all agree that scientists must engage in debate that is non-scientific in this sense. However, Collins's critics will (correctly, I think) regard such a definition of non-scientific as humpty-dumptyish and will see nothing wrong with regarding much argument and evidence as "scientific" even when it is not rule-governed or algorithmic.

Since it seems that Collins and his critics can be brought to the bargaining table, we can now raise the question of whether a model of science acceptable to both parties might be formulated. I turn to

# Beyond the Science Wars

Richard Bernstein's enlightening discussion of scientific rationality in his *Beyond Objectivism and Relativism: Science, Hermeneutics, and Praxis* (Bernstein 1983). Bernstein notes that the early reactions of philosophers of science to Thomas Kuhn's *The Structure of Scientific Revolutions* were harsh (p. 51; see also Lakatos and Musgrave 1970). Kuhn was taken as advocating irrationalism and subjectivism. We saw in chapter 3 that there are good grounds for such an interpretation of the early Kuhn. However, Bernstein thinks that a sympathetic reading of Kuhn, particularly of his later writings, shows that he was not (or did not long remain) an irrationalist or subjectivist (p. 53). For instance, he quotes the following passage from the postscript (written in 1969) to the second edition of *The Structure of Scientific Revolutions*:

> Nothing about that relatively familiar thesis [that theory choice is not a matter of deductive proof] implies either that there are no good reasons for being persuaded or that those reasons are not ultimately decisive for the group. Nor does it even imply that the reasons for choice are different from those usually listed by philosophers of science: accuracy, simplicity, fruitfulness, and the like. What it should suggest, however, is that such reasons function as values and that they can thus be differently applied, individually and collectively, by men who concur in honoring them. If two men disagree, for example, about the relative fruitfulness of their theories, or if they agree about that but disagree about the relative importance of fruitfulness and, say, scope in reaching a choice, neither can be convicted of a mistake. Nor is either being unscientific. *There is no neutral algorithm for theory-choice, no systematic decision procedure which, properly applied, must lead each individual in the group to the same decision.* (Kuhn 1970, pp. 199–200; emphasis added)

Bernstein interprets this and other passages as indicating that Kuhn conceives of the rationality exhibited by scientists in their disputes about theories as very similar to the kind of reasoning Aristotle associates with *phronesis*:

> Many of the features of the type of rationality that is exhibited in such disputes show an affinity with the characteristics of *phronesis* (practical reasoning) that Aristotle describes. Aristotle, of course, was not addressing the problem of disputes about rival paradigm theories, and in his analysis of *phronesis* he contrasts it with *episteme* . . . as well as *techne*. But *phronesis* is a form of reasoning that is concerned with choice and involves deliberation. It deals with that

which is variable and about which there can be differing opinions (*doxai*). It deals with a type of reasoning in which there is a mediation between general principles and a concrete particular situation that requires choice and decision. In forming such a judgment there are no determinate technical rules by which a particular can simply be subsumed under that which is general or universal. What is required is an interpretation and specification of universals that are appropriate to this particular situation. (Bernstein 1983, p. 54)

The parallels between Kuhn's (at least, the later Kuhn's) view of the process of theory choice and the deliberations of Aristotle's practically wise man are striking. These parallels are made even clearer in Kuhn's essay "Objectivity, Value Judgment, and Theory Choice" (Kuhn 1977, pp. 320–39). For Aristotle the practically wise person knows the good and can deliberate well on the best means of attaining it (see the *Nicomachean Ethics,* 1140a26–29). For Kuhn "the good" in situations of theory choice will be such qualities of theories as accuracy, simplicity, consistency, scope, and fruitfulness (Kuhn 1977, pp. 321–22). The good scientist is one who can deliberate well about which theories will best instantiate those values.

Aristotle says that since the means for attaining the good are contingent, demonstration is not the sort of reasoning employed in the practically wise person's deliberations (1140a32–35). Similarly, Kuhn insists (in agreement with Collins) that no algorithm or decision procedure can specify the results of our theory-choice deliberations (Kuhn 1977, p. 326). Nevertheless, as interpreted by Bernstein, Kuhn's mature view was that theory change is an eminently *rational* process (Bernstein 1983, p. 56). Theory choice is preceded by a period of reasoned debate and dialogue between the qualified parties. The debate is guided throughout by shared values and standards; the judgments and deliberations involved in theory choice are always "eminently discussible" (Kuhn 1977, p. 337). Further, as interpreted by Bernstein, the notion of incommensurability seems to have completely disappeared from Kuhn's later writings.

I am not sure whether to endorse Bernstein's interpretations of Kuhn and Aristotle. To me, Kuhn really does evince a volte-face between 1962, when the first edition of *The Structure of Scientific Revolutions* came out, and 1969, when the postscript was written for the second edition. Whether Bernstein's interpretations of Kuhn or Aristotle are entirely accurate is beside the point. The point is that Bernstein has outlined a rationalist, nonalgorithmic, and very plausible model of theory choice, which I shall call the "dialogical" model in opposition to Collins's algorithmic model. According to the dialogical model, theory choice is rational because it is brought about through a process of reasoned dialogue and debate between qualified discussants who draw upon broadly shared standards and values and a vast amount of deeply grounded background beliefs about theories, facts, methods, etc. At no point will any sort of incommensurability be encountered, nor will there be any need to resort to rhetoric, threats, cajolery, or disguised appeals to social interests (though, as we have seen, such elements are unfortunately often part of scientific debate). Even debates about fundamental values, for instance, accuracy versus simplicity, can be rationally addressed—philosophers do so all the time.

The dialogical model emphasizes that no algorithm can determine theory choice. The debates over theory choice involve the application of practically learned skills of judgment and deliberation within concrete dialectical circumstances. Further, the dialogical model of theory choice fits right in with Marcello Pera's dialectical view of scientific rationality, which we examined in chapter 4. Pera would certainly agree with Collins that science does not involve the algorithmic application of universal rules of method. As we saw, Pera holds that the rationality of science lies in the fact that scientists prefer that theory which is supported by the strongest arguments (Pera 1994, p. 144). The dialogical model of theory choice developed in this chapter simply fleshes out the process whereby scientific communities decide which arguments are the strongest: reasoned debate involving the application of practically learned judgmental and deliberative skills in the context of a vast amount of shared background beliefs, values, methods, etc.

How is it known when an issue has been debated enough and victory for one side should be declared? Here again, no automatic decision procedure dictates the decision. It is a matter of the informed *collective* judgment of a scientific community. However, this does not mean that favorable judgments must be bought with disguised appeals to social interests. As Toulmin observed in the quotation at the end of chapter 3, by the time the heliocentric theory had won, the overwhelming arguments and evidence were *there*. Nobody had to be cajoled, manipulated, threatened, deceived, or bought off with appeals to social interests (Toulmin 1972, p. 105). Occasional intrusion of external influences (as by the Pope in the Galileo case) had little effect in the long run. Despite Galileo's condemnation in 1633, the Copernican Revolution was essentially complete by 1650 (Ashworth 1997).

The upshot is that there is an alternative *rational* model to Collins's algorithmic account. According to this alternative, shared and deeply grounded values, standards, and knowledge guide but do not determine the outcome of scientific debate. Closure is achieved through rational debate in which persuasion involves the application of practically learned skills of judgment and deliberation. The fact that the decision to declare a debate closed must involve collective judgment does not mean that the decision was arbitrary or had to be motivated by tacit appeals to social interest. Copious evidence and argument will back these judgments. That such evidence might not be sufficient to *compel* but only to *persuade* is not a weakness of the model.

The dialogical model of science can be tested vis-à-vis the dinosaur extinction controversy. Among the predictions of that model are that even the deepest scientific disagreements, those involving gut-level commitments, admit of rational resolution. Scientists viscerally opposed to new theories can be rationally persuaded to abandon their enmity, or at least to soften their opposition enough to give the new theory a chance. Further, they can be persuaded to do so without undergoing semireligious conversions, or being browbeaten, inveigled, or bribed into acquiescence.

Scientists are least appealing when they dismiss a new theory without a hearing. Perhaps they do listen but respond with harsh polemics and the clear intention of smashing the theory and disgracing its supporters. Through the ages scientific mavericks (and crackpots) have consoled themselves with the thought "They laughed at Galileo, they laughed at Newton, they laughed at Pasteur. . . . " Of course, as Martin Gardner notes, they also laughed at Bozo the Clown, and Bozos have always outnumbered Galileos ten thousand to one (at least). Still, many of the greatest scientists have had to endure the contempt of colleagues. Darwin's letters are especially poignant when he mourns the truculent responses of old friends and teachers to *The Origin of Species*.

The sometimes vicious opposition faced by new theories is often viewed as the dark side of science (recall Raup's complaint against "philosophical" prejudice; Raup 1986, pp. 197–201). Of course, scientists are just as prone to prejudice as are, say, sociologists of knowledge, but not all of the automatic opposition to radical new ideas is due to bias. As noted in chapter 3, the human imagination produces far more hypotheses than it is practical to test, so there must be criteria to choose the best candidates for deeper investigation. These criteria are not always expressions of prejudice, but usually reflect the practical experience of scientists who have suffered years of frustration and time wasted in conceptual cul-de-sacs. Practically learned judgments guide the scientist out of blind alleys or away from the pursuit of ignes fatui.

For over a century and a half, the principle guiding geological theorizing has been actualism, which I defined in chapter 3 as the principle that the hypothesized causes of geological phenomena should not differ in kind from those actually observed. I defined "uniformitarianism" as adding the further stipulation that such known causes not be postulated to have acted with a degree or intensity never observed. As noted in chapter 3, the earth sciences abandoned such strict uniformitarianism some time ago. In fact, in the K/T extinction debates both impactors and volcanists have postulated events of vastly greater magnitude than anything ever observed by humans.

According to William Glen, some volcanists have moved well beyond solid observational and experimental evidence in speculating that explosive volcanism of unprecedented intensity could account for the production of shocked minerals (Glen 1994, pp. 31–32). Charles Officer, perhaps the staunchest opponent of impact theories, attributes the K/T extinction largely to the massive volcanism that caused the Deccan Traps formation in the Indian subcontinent (Officer and Page 1997, p. 170). He admits that no such extensive volcanism has occurred during the last 65 million years, so clearly humans cannot have observed volcanism even approaching such intensity (p. 165).

Still, Raup may have been correct that, as late as 1980, a "philosophical" bias against catastrophism lingered in the earth sciences. The distinction between differences of kind and differences of magnitude is not absolute. Extreme differences in degree are often tantamount to differences in kind. The Alvarez hypothesis postulated an impact many orders of magnitude greater than even the most destructive ever witnessed (e.g., Tunguska). Further, the accepted view was that truly large impacts had occurred only during the earliest period of earth history. The impact theory may therefore have seemed like a deus ex machina—a speculative scenario postulating a causal agent of such an extreme degree that it amounted to a difference in kind from known causes.

In July 1994 the most closely scrutinized astronomical event of all time, the crash of comet Shoemaker-Levy 9 into Jupiter, showed that truly massive impacts do presently occur in the solar system. The larger fragments of the comet were estimated to range from 0.5 to 3 km in size and were traveling at 60 km/s (Weissman 1995). Had even one hit the earth, the results would have been cataclysmic. By the early 1990s astronomers had estimated that twenty-one hundred asteroids of greater than 0.9 km diameter cross the earth's orbit (Gaffey 1992). Equally remarkably, in 1990 scientists reported the discovery of the "smoking gun"—the actual crater from the impact said to have caused the K/T extinctions (Glen 1994, pp. 14–18). They claimed that the so-called Chicxulub crater off the Yucatan is 200 km wide and is dated precisely to the K/T boundary (see Frankel 1999 for an updated report on the Chicxulub claims).[7]

It follows that the observational basis for impact hypotheses is much stronger now than in 1980. In 1980 talk of a bolide impact causing the K/T extinctions may have seemed like the invocation of a deus ex machina. Given the evidence for massive bolide impacts gained since then, such a judgment now seems much too harsh. If anticatastrophist bias formerly inclined people against the impact theory, such prejudice has surely waned. Lyell himself, after witnessing the Shoemaker-Levy impacts, could hardly proscribe such theories.[8]

Finally, the impact hypothesis does not have to be swallowed whole. Conservative earth scientists might come to accept part of that hypothesis, for instance, that Earth did suffer one or a series of bolide impacts at about the time of the K/T extinctions, but reject other parts, such as that these impacts were the main cause of the mass extinctions. This is precisely what happened. For instance, by 1989 prominent paleontologist Antoni Hoffman had accepted that studies of shocked quartz from the K/T boundary show that these grains could not have been produced volcanically. This and other evidence, such as apparent tsunami deposits at K/T boundary sites, strongly suggested to him that a bolide impact did take place at the end of the Cretaceous. He continued to hold that the complexity of the fossil record rules out any single-cause extinction theory and instead indicates the coincidental occurrence of two or more deleterious factors

## Beyond the Science Wars

(Hoffman 1989b, p. 9). Aspects of the impact theory judged amenable to accepted standards might therefore be adopted and other aspects repudiated (Archibald 1996 advocates an updated version of such a synthetic view).

The upshot of my argument here is that earth scientists are not confronted with the stark choice between one monolithic set of theory-cum-standards and another such set. As noted in chapter 3, the impact theory largely appeals to conservative standards. In Raup's particular case, the theory was adopted in response to a new reading of the fossil record, that is, as evincing periodic extinctions that no earthly process could explain. The "philosophical" barriers between impactors and gradualists—if they really existed—have been breached by *empirical* evidence acquired since 1980. Finally, a synthesis can be made from bits and pieces of both theories, with those parts retained that are thought conformable with the best evidence.

So, as the dialogical model of science predicts, even the most virulent, visceral opposition to new theories can be overcome, or at least greatly weakened, by rational argument and new observations or discoveries. Compromises and common ground can be found even with highly antithetical claims.[9] No incommensurability is encountered, the contending parties need not experience "born again" sorts of conversions, and there is no need to buy off opponents by appeal to their vested social interests.

Maybe then there is hope for lasting peace in the science wars. Maybe the Kuhn-Bernstein view of theory choice, what I have called the "dialogical" view of science, can provide the *via media* between scientism and constructivism. Zealots on both sides of the science wars may be unhappy with this proposal. For those seeking an end to the science wars, the dialogical model is an attractive option. It agrees with Collins that no algorithm can settle scientific debates and that these debates are complex and contextual. However, it also supports the rationalists' insistence that such debate is eminently reasonable and is guided throughout by rational principles. Instead of a robotized dispenser of algorithms, the scientist becomes the practically wise deliberator about how objective values are best instantiated in a concrete dialectical situation.

The science wars will end when the unexciting, but true, view prevails: Science is not simply a mirror held up to nature; it is much more complex and interesting than that. No algorithm can specify theory choice; it is a matter of collective decision involving rational debate and the application of practical wisdom gained through long experience. So, as the constructivists have insisted, all theory-choice decisions will involve negotiation and, inevitably (scientists being merely human), all the devices of rhetoric and the ploys of politics. Yet for all that, there is no reason to deny the claim of traditional rationalists that scientific theory is rigorously constrained by our observations of and interactions with the physical world. Further, scientists *can* be persuaded by traditionally recognized reasons, and there is no basis for the cynical conclusion that they almost never are. The circle of the "experimenter's regress" can be broken by rational argument; appeals to vested social interests are not necessary. Finally, there is no reason to regard the standards or methods of science as arbitrary "rules of the game."

★ ★ ★

Where does all this leave dinosaurs? Can we draw out Leviathan and say what dinosaurs were *really* like? Yes and no. We do know some things about dinosaurs and are learning more all the time. Some things we shall never know. The colors of dinosaurs will remain speculative. Were they the gray-green that got Mitchell's interpretive juices flowing? Were they purple, or pink with green polka dots? No one will ever know.

What are the prospects of creating a Jurassic Park, of bringing the dinosaurs back? Pretty dim, say Rob De Salle and David Lindley, a paleontologist and physicist who coauthored a popular book on the issue (De Salle and Lindley 1997). Even if a Jurassic Park were feasible, it would be only a zoo, not a restoration of the dinosaurs' world. Many details of the appearance, physiology, behavior, ecology, and extinction of dinosaurs will therefore remain matters of speculation. We hardly know everything about living animals, so how can we completely understand fossil ones?

Someone once observed "Theories change. The bullfrog remains." So *Tyrannosaurus rex* remains, burning bright in the forests of the Cretaceous night, safely extinct in the depths of time, long, long before our theories could frame its fearful symmetry. Knowledge begins with wonder, and that is the great value of dinosaurs for us, the primal wonder inspired by knowing that such beings once actually walked on this planet. Since we love them, we would like to know everything about them, but are frustrated by the myriad centuries separating us. When we think about dinosaurs, therefore, imagination must always supplement fact. With creatures as wonderful as dinosaurs, this is the way it should be.

# NOTES

## 3. The "Conversion" of David Raup

1. The Alvarez team was not offering an explanation for all mass extinctions, only the one at the K/T boundary. The impact theorists' claim has always been that bolide impacts were sufficient but not necessary for mass extinction. However, according to Frankel (1999), evidence is mounting for impact hypotheses accounting for other mass extinctions, including the Permian extinctions, the biggest of all.

2. This dichotomy between bolide and volcanic theories is too simple. As Glen notes, some scientists hold that both volcanic and impact events could have contributed to the K/T extinctions (Glen 1994, p. 73). The point is better expressed by saying that only the volcanic or impact theories, or some combination of the two, can possibly account for the shocked quartz, iridium layer, etc.

3. Like so much else in Kuhn's work, his view of conversions is hard to understand clearly. Kuhn himself, and his more sympathetic commentators, have denied that anything irrational is implied by his talk about "conversions" or "gestalt switches." For instance, Paul Hoyningen-Huene (1993, p. 258) argues that Kuhn uses the term "conversion" to emphasize that belief changes are involuntary, not that they are irrational. Hoyningen-Huene insists that for Kuhn revolutionary change is always explicable in terms of "identifiable reasons."

It may therefore seem unfair that I have associated Kuhn with an antirationalist view of scientific change. Yet I think it is undeniable that the overwhelming majority of those who have read and reacted to Kuhn, both foes and would-be friends, have taken him to say something radical about theory change. It may well be that those readings, as Hoyningen-Huene insists, are based on misunderstanding. However, when the misunderstanding has carried the day, we have no choice but to deal with it. Therefore, I would like readers to understand that when I attack "Kuhn" in this chapter, I am attacking a particular, widespread *interpretation* of his view. Perhaps I am attacking a straw man, but, if so, the straw man has grown so large that he eclipses the real one.

Why has Kuhn been so widely misunderstood, if he has? Perhaps he is a victim of the famous Murphy's Law that states "When you speak so clearly that nobody can misunderstand you, you will be misunderstood." Actually, I think Kuhn's own language about "conversions," "world-changes," "gestalt switches," etc., naturally led to such (mis)interpretations. It is undeniable that many sociologists of knowledge saw Kuhn as taking the study of science away from philosophers, who insisted on interpreting theory change in terms of *reasons,* and handing it to sociologists, who gladly viewed scientific change as an effect of social causes. Sociologist Barry Barnes, whom I quote in this chapter, affirms precisely this. Robert Klee summarizes the impact of Kuhn:

> The demise of the positivist model of science and the rise of Kuhnian historicism ushered in a momentous change in philosophy of science. The previous consensus view that science, at least on occasion, discovered a preexisting objective reality gave way to alternative critical accounts of science that stripped it of any such achievement. These new critical accounts were invariably relativist in epistemology, and, for the most part, antirealist in ontology.... The new forms of relativism and antirealism were born within the nest of the social sciences. Many Kuhn-inspired critics of science took Kuhn's ultimate point to be that the philosophy of science, as traditionally conceived since the time of Immanuel Kant, was dead. Its old job would be taken over by a successor discipline called variously the sociology of science, the sociology of knowledge, or science and technology studies. (Klee 1997, p. 157)

4. Harold Rollins points out (personal communication) that stratigraphers are aware of other "pencil-thin" marker horizons such as volcanic ash beds, so perhaps Glen is overstating his case here. However, Glen's point is that other such markers, such ash beds, are with rare exceptions limited only to parts of single continents (Glen 1994, p. 78). Glen claims that the iridium layer is the first truly *global* such marker, and that this is what makes it so remarkable (p. 79).

5. The perfect gloss on the (mis)appropriation of Kuhn by sociologists and others is given by Robert Klee:

> The publication of Kuhn's *The Structure of Scientific Revolutions* constituted one of those rarest of events in philosophy: a philosophical work with an even larger impact outside professional academic philosophy than inside.... Kuhn's general model of systematic inquiry struck a chord throughout the intellectual community, from social scientists to literary critics to poets and writers and physical scientists. There are many professional intellectuals in nonphilosophical disciplines who know virtually no philosophy—except that they know (or think that they know) Kuhn's views about science. In short, Kuhnian philosophy of science became instantly faddish and experienced the fate of every victim of a fad: distortion and misuse. Everybody and their aunt and uncle felt free to cite Kuhn, to claim that Kuhn's views supported whatever their own pet position was, and to argue for their views by presupposing Kuhnian principles, as though the truth of those principles could hardly be doubted by any bright and well-educated person. (Klee 1997, p. 130)

6. Harold Rollins suggests that I may draw too stark a contrast between impactors and more traditional earth scientists. He tells me (personal communication) that he and other geologists did not initially perceive quite so sharp a dichotomy between the two camps. He suggests that I point out that my (and Raup's) distinction between the two has heuristic value for setting up a desired contrast, but that the attitudes of geologists at the time cannot be so simply characterized.

Professor Rollins also points out that one chief difference between proponents of the Alvarez view and more traditional earth scientists is that the former were seeking a single cause of the K/T extinctions. The latter group, perhaps more used to viewing such events as complexly caused, preferred to explore the issue with multiple working hypotheses.

7. The following is from a personal communication from Raup:

Let's suppose that you are correct in your analysis and ask: "Why did he [Raup] say those things about his conversion?" I can think of some possible reasons. First, in *Nemesis Affair*, I tried to avoid criticizing colleagues, and used myself as a proxy. And second, I have always enjoyed putting myself down in public—or at least, not putting myself up—because the [sic] life is more challenging (and fun) if one starts from a weak position. I admit to practicing "self-reproach that is out of proportion to the sins actually committed."

8. Robert Olby points out (personal communication) that referee reports often start nice and then get nasty. Thus, since I have only the first page of Raup's report, I cannot tell how nasty it eventually gets. This is a fair criticism, but I stand by my claim that Raup's early views were not as inimical to catastrophic extinction scenarios as *The Nemesis Affair* would lead us to believe. Raup responds to this claim as follows: "You suggest that I was never really all that committed to actualism. This is probably true and surely stems from the influence of my botanist-forester father. He was a strong iconoclast on most things and his research pioneered the recognition of natural disturbance (wind, fire, etc.) as an important force in plant communities" (Raup, personal communication).

9. To accept that a bolide impact occurred at the end of the Cretaceous does not per se commit one to the view that the iridium layer is the definitive marker of the K/T boundary. One could still hold that the effects of the impact played out over geological time. In this case, the K/T boundary would be defined (as it traditionally was) in terms of faunal change. However, the view of Alvarez et al. and of Raup himself was that the extinctions were truly sudden and catastrophic (Raup 1991). On this view, the iridium layer does serve as a marker for abrupt global change.

10. The earlier deterministic theories which Raup is opposing with his model (which incorporates both stochastic and deterministic explanations of mass extinctions) all assumed that mass extinctions would be caused by unique, that is, nonrecurring, events, such as extreme climatic cooling (where such cooling was not part of periodic warming and cooling processes such as the Milankovich cycle). Raup at this stage simply retains this assumption of nonrecurrence with respect to deterministic explanations of mass extinctions. Only later does his work with Sepkoski lead him to postulate a periodic cause of mass extinctions.

11. It might be argued that while neutral techniques may be available in a controversy, the decision to use such a technique, or to trust such a technique over other available ones, may not be neutral. As I note, not all qualified parties accepted Raup and Sepkoski's mathematical arguments for periodicity. Antoni Hoffman, for instance (who rejected extraterrestrial explanations of mass extinction), regarded the extinction data as compatible with a random walk (Hoffman 1989a). However, I am not arguing that the availability of neutral techniques precludes the need for interpretation of data or that theoretical commitments do not influence scientists' choices of techniques.

In this chapter I am making the very minimal claim, *pace* the radical Glen-Kuhn thesis, that neutral standards, values, techniques, methods, etc., *do exist* and *are employed* by the conflicting parties in scientific controversies. Further, I am suggesting that the mathematical analyses and tests performed by Raup and Sepkoski might have been sufficient to *rationally persuade* them of the reality of periodic mass extinction. I am not claiming that these analyses and techniques were cogent enough to *compel* the assent of doubters like Hoffman. I interpret the early Kuhn (an interpretation I share with Lakatos 1970, Kordig 1971, Newton-Smith 1981, and Scheffler 1982, among others) as making the radical claim that proponents of different theories are hermetically sealed in conflicting *Weltanschauungen* with no possibility of adducing neutral grounds sufficient to persuade their opponents. This is the "exciting" interpretation of Glen's data, which I am addressing in this chapter. It is this radical claim that I am opposing; I am not offering an equally radical claim in the other direction.

12. I sent Prof. Raup a draft of this chapter. Here is his response to my conclusion:

> If there was a Kuhnian conversion in my case, it was surely a slow one. Also, the emphasis was on the catastrophism—gradualism (or actualism) issue, rather than on the bolides themselves. And I think there *was* an involuntary, viscerally controlled world view that was converted, over the course of a few years, to another world view. I happily concede that the "conversion" was based on rational treatment of real data. But, I submit, when I (or anyone) is entrenched in one or the other world view, the viscera takes over rationality in evaluation of all but the most compelling new data. I think the conversion in my case took a lot longer than it should have—or would have if I had not been "controlled" by the Lyell-Darwin model. (Raup, personal communication; emphasis in original)

As I read this passage, Raup seems to agree with me that his "conversion" was very different from the Glen-Kuhn sort. Raup says that his change was based on "rational treatment of real data" and took a few years to complete. Further, the rationalist need not deny that global "involuntary, viscerally controlled" changes of perspective do occur. If these changes follow years of rational treatment of real data, as Raup says was the case with him, the conclusion seems to be that these involuntary changes in perspective were ultimately driven by *reasons*. It is not the *magnitude* of a conceptual change that concerns the rationalist, but *how* that change came about. Was it driven by reasons or not?

## 4. Are Dinosaurs Social Constructs?

1. I have phrased my characterization of the rationalist view so as not to presuppose scientific realism. This is because I do not wish to exclude all antirealists from the rationalist camp. Some harsh critics of scientific realism, like Larry Laudan, are equally astringent critics of constructivism and relativism (Laudan 1990b, 1996). Antirealists can certainly be rationalists on my definition. Even if the goal of science is to achieve empirical adequacy rather than accurate representation (van Fraassen 1980), it is how we observe *nature* to be that tests claims

about a theory's empirical adequacy. The constructive empiricism of Bas van Fraassen assumes a clear distinction between theory and observation (van Fraassen 1980); that is, there must be a realm of theory-neutral observational facts. The test of a theory's empirical adequacy is how well its predictions stand up to how we *observe* nature to be. Empirical adequacy is therefore no more a social construct than representational accuracy.

2. I sometimes wonder if Latour and Woolgar's tortuous prose is not itself a rhetorical device. Many passages seem needlessly obscure. I once heard someone accuse Jacques Derrida of the tactic of "obscurantist terrorism." This is the practice of writing so opaquely that when criticized one can disdainfully disown any critic's interpretation. Perhaps Latour and Woolgar had this tactic in mind. On the other hand, they specifically endorse those schools of literary criticism that deny that texts *have* meanings other than whatever meaning readers attribute to them (Latour and Woolgar 1986, p. 273). In this case, it will be impossible to *misinterpret* them, so they cannot claim that I have misread them here. In short, I have their explicit permission to interpret them any way I please.

3. One reader of this chapter felt that I had been far too harsh on Latour and given far too pejorative an explication of his views. He argued that Latour was merely explicating the nature of scientific rationality. I am tempted to respond flippantly by asking what Latour's characterization of scientific *irrationality* would look like. More to the point, Latour himself is quite explicit in calling for a moratorium on cognitive explanations of science (Latour 1987, p. 247). How Latour could explicate scientific rationality while banning any appeal to cognitive factors is an utter mystery to me.

Latour further rebukes the Enlightenment's dreams of reason in the "Irreductions" in *The Pasteurization of France*. Here is a sampling:

2.1.8.2. No set of sentences is by itself either consistent or inconsistent . . . all that we need to know is who tests it with which allies and for how long. Consistency is felt; it is not a diploma, a medal, or a trademark. (p. 179)

2.2.2. Since nothing is reducible or irreducible to anything else . . . and there are no equivalences . . . , every pair of words may be said to be identical or to have nothing in common. Thus, there are no clear ways of distinguishing literal from *figurative* meanings. . . . Every group of words may be dirty, exact, metaphorical, allegorical, technical, correct, or far fetched. (p. 181; emphasis in original)

2.3.4. Nothing is by itself either logical or illogical . . . , but not everything is equally convincing. There is only one rule: "anything goes"; say anything as long as those being talked to are convinced. You say that to get from B to C, you have to pass through D and E? If no others raise their voice to suggest other ways, then you have been convincing. They go from B to C along the suggested path even though no one wants to leave B for C and there are lots of different routes that could have been taken. Those you sought to convince have acquiesced. For them, there is no more "Anything goes." That will have to do, *for you will never do any better.* (p. 182; emphasis in original)

2.4.3. We cannot distinguish between those moments when we have might and those when we are right. (p. 183)

2.5.4. We neither think nor reason. Rather, we *work* on fragile materials—texts, inscriptions, traces, or paints—with other people. These materials are associated or dissociated by courage and effort; they have no meaning, value, or coherence outside the narrow network that holds them together. (p. 186; emphasis in original)

Whatever Latour intends by these oracular utterances, it is clear that he repudiates what has so far passed for scientific rationality (not to mention elementary logic).

4. Readers friendlier to constructivist theses will still no doubt hold that I have unfairly characterized Latour in the harshest and most unfavorable manner. One reader of this chapter accused me of engaging in "mudslinging" and encouraging people to dismiss Latour. Allow me to take this opportunity to exhort readers unfamiliar with the works of Latour not to take my word for it. I strongly urge them to read the works I cite here and judge for themselves if I have been too harsh.

5. Just as this book was being finished, the book *Scientific Controversies: Philosophical and Historical Perspectives,* edited by Peter Machamer, Marcello Pera, and Aristides Baltas, was published by Oxford University Press (2000). This book presents a number of case studies and philosophical analyses that parallel my treatment of scientific controversy. The authors present copious evidence against constructivism and for the *rational* resolution of scientific controversies.

6. A standard maneuver of skeptics is to claim that they are only pointing out the skeptical consequences of their opponents' premises. Latour therefore might try to construct a reductio ad absurdum, employing only premises accepted by rationalists, showing that self-contradiction follows from the assumption (what I called assumption C) that some arguments have intrinsic logical or evidential cogency, whether or not they actually persuade. Latour's reductio ad absurdum (LRAA) would have the following form:

$p_1$: C           (assumed premise)
$p_2$
$p_3$
.
.
.
$p_n$: q and non-q           (derived contradiction)

Conclusion: not-C,

where $p_2$ through $p_n$ would be only premises accepted by rationalists and logical consequences of those premises.

The conclusion not-C is in the form of a universal generalization: "No argument has intrinsic logical or evidential cogency." Since this is a universal generalization, it has LRAA itself as one of its instances. In other words, if not-C is true, no argument, not even LRAA, can provide any grounds for its truth.

7. In his recent book *The Last Word,* Nagel shows again and again that every attempt to debunk objectivity inevitably founders when it makes, as it must, the very sort of objective claim it seeks to deny:

sider the ongoing debates between Christian apologists and skeptics over the purported resurrection of Jesus (see, e.g., Martin 1998, 1999; Davis 1999). It seems highly unlikely that any new documentary evidence will turn up pertaining to this issue. Also, it seems improbable that any remarkable new methodological advances will help resolve things. Nevertheless, apologists and skeptics can promote their theories as the most reasonable account of the data. For instance, William Lane Craig (1994) argues that the reports of post-crucifixion appearances of Jesus are best explained in terms of a literal resurrection from the dead. Gerd Ludemann (1995) disagrees and contends that all of the so-called appearances can be explained as visions.

In the natural sciences, instead of working to produce the best account of a relatively static body of evidence, scientists have to try to keep up with a torrent of new data and a stream of methodological and technological innovations. Often theorists are kept running just trying to accommodate the influx of data; see Louie Psihoyos's essay on recent dinosaur discoveries in Currie and Padian (1997). The result is that when new scientific theories are accepted, they are often supported by vastly more and better data than were the old theories they replace.

4. The fruitful interaction of theory and practice in progressive episodes is illustrated by the acceptance of continental drift by geologists during the 1960s. A. Hallam describes the development of this hypothesis and its reception among professional geologists:

> The highly successful modern theory of plate tectonics, which has an all-pervasive influence on the earth sciences today, may be regarded as an outcome and development of the continental-drift hypothesis, which provoked intense controversy for half a century following its presentation by Alfred Wegener. For many years its adherents were often dismissed contemptuously as cranks by the geological establishment on both sides of the Atlantic, though more especially in North America. At best the idea was considered inadequately supported by evidence and mechanically implausible, and it had no serious effect on the mainstream interests of most earth scientists. (Hallam 1989, p. 135)

What the hypothesis needed was definitive evidence that drift had occurred and a plausible theory of the mechanical means of continental movement.

The evidence began to accumulate with the development of new techniques: "After the Second World War the application of new or more refined geophysical techniques began to transform our knowledge of the earth in a radical way, and these were eventually to lead to a profound revolution in thought from a stabilist to a mobilist view" (Hallam 1983, p. 162). For instance, the development of a new and much more sensitive magnetometer permitted far more extensive studies of geomagnetism (p. 163; the declassification of U.S. Navy magnetometer studies also helped). The best evidence concerned the spreading of seafloor from midocean ridges, structures which were not discovered until the 1950s when new echo-sounding equipment made such topographic surveys possible (p. 165). For many geologists, decisive confirmation of seafloor spreading came with the confluence of several new lines of evidence generated by various new or improved methods and techniques (pp. 171–72).

What was still needed was a theory of how it is mechanically possible for continents and seafloor to move. This was provided by the theory of plate tectonics, first suggested by Tuzo Wilson in 1965. As developed by Jason Morgan, this theory proposed that the earth's crust is part of a rigid, less dense layer of rock, later called the "lithosphere," which rests on a denser, less rigid "asthenosphere" (p. 173). The earth's crust is thus composed of a number of rigid plates, which can move because they float on denser but more plastic material.

The drift hypothesis, which was rightly rejected for fifty years, was ultimately confirmed by the confluence of new ideas and new practices. New ideas stimulated the development of new techniques, and new techniques permitted the more thorough testing of new ideas. Technological advances independently developed were appropriated by geophysicists to produce more and better data. These data in turn stimulated new hypotheses. The confirmation of the drift hypothesis exemplifies the complex interaction of theory and method that characterizes scientific progress. The upshot is that scientific progress involves advances in both theory and practice: Successive theories must evince greater accuracy, empirical adequacy, etc., and science must be *done* more efficiently, thoroughly, reliably, etc.

5. R. A. Thulborn argues that Alexander employs a simplistic method of estimating dinosaur leg height from foot length (Thulborn 1989, pp. 41–42). Thulborn recommends that estimates instead be based on allometric equations that more realistically reflect vertebrate growth patterns (p. 42).

6. Shapin and Schaffer create a straw man when they imply that the opposing tradition viewed Boyle's triumph as "self-evident" or "inevitable." I view it as neither. To say that Boyle and his defenders had the better case is not to say that Hobbes was so obtuse as to deny self-evident truths. We merely judge that, in the end, Hobbes was wrong and that the application of the experimental method was rationally justified. Also, Boyle's success was certainly not inevitable: Had the Armada succeeded in 1588, Catholicism (and the Inquisition) would have been established in Britain, and it is most unlikely that there would have been a Royal Society or any air-pump experiments.

7. The weakness of Feyerabend's case is also shown by comparing methodological rules with moral ones. In general, lying is wrong, but in warfare it is proverbial that truth is the first casualty. Winning the Second World War required massive campaigns of deception and disinformation, such as the creation of whole phantom armies to deceive the German high command about the D-Day landings. Clearly, there are times when the consequences of telling the truth are so bad that lying and deception become permissible, even obligatory. Likewise, almost any methodological rule can occasionally be overruled by more pressing obligations (epistemic or moral).

Methodological norms, like ethical ones, are therefore generally prima facie rather than absolute (though some obligations, like avoiding self-contradiction, seem to hold good in every situation). The fact that the prima facie duty not to lie can be occasionally overridden does not license mendacity. Likewise, even if some methodological norms can be justifiably ignored on occasion, it does not

## Notes to page 144

follow that anything goes. Nihilism does not follow from the occasional exceptions to ethical norms and epistemological anarchy is not warranted just because methodological prescriptions are sometimes overruled.

8. Some philosophers are still haunted by the fear that a fixed, universal, largest-scale Method is the only alternative to relativism (Worrall 1988). That this need not be so is cogently argued by Alan Chalmers (1990, 1999). Chalmers shows that relativism can be resisted by appeal to low-level, nonuniversal methods, such as Galileo's telescopic techniques (Chalmers 1990, pp. 54–60). Of course, anyone advocating such a new method must give good reasons for its acceptance, and naturally those reasons will appeal to other methods and standards. Cannot relativistic skepticism then be aimed at those further methods and standards?

The problem with relativism is not that it seeks to criticize any *particular* standard or method. Particular methods or standards can be rejected, modified, or overruled at one time or another, but it is simply incoherent to say that *all* can be simultaneously repudiated. Further, relativism, incongruously, is just as universal, aprioristic, and ahistorical as the most egregious positivist methodism. When relativists declare that all methods and standards are equal, they presume to speak from a standpoint beyond history; they apparently intend to voice an eternal epistemological truth. No historically situated inquirer—one actively involved in a process of intellectual inquiry—ever would or could say such a thing, upon pain of self-refutation. In real life we have *no choice* but to employ the standards and methods that seem right to us.

Relativists might reply that they are simply performing an inductive generalization: Historically, every set of epistemic norms has eventually been supplanted by other, incompatible norms. Therefore, it is a very good bet that none of our present norms is absolute, and all will someday be superseded. Of course, this relativist argument itself invokes a supposedly correct rational procedure—inductive generalization. But in offering this argument relativists need not *endorse* inductive reasoning; they use it because rationalists *do*. They thereby intend to show an internal inconsistency of rationalism, that is, that one of the rationalist's own methods leads to the conclusion that there can be no universal norms.

Rationalists can respond that from their perspective there have been many instances in many fields of inquiry where long periods of failure were followed by success. For instance, the transmission of genetic information was not understood until the discovery of DNA. Mathematical proofs sometimes elude mathematicians for centuries before they are finally formulated. Therefore, unless from Aristotle on there has been no or very little progress in elucidating scientific methods, and relativists have not shown this, there is no warrant for concluding that the search for Scientific Method must ultimately fail. Maybe it is just a very difficult task.

In fact, though, as I argue in this chapter, when new methods are adopted by a given science, it is often a change for the *better*. So, *pace* the relativists' intentions, the application of inductive generalization to the history of science seems to show that insofar as methods do change, they tend to change for the better. Much scientific progress consists of finding better ways to do science, and we should inductively generalize that such progress will continue. But, of course, the evaluation that an episode of method change was an improvement implies the existence of standards for judging such transitions. Are there such standards? I think that Nagel (1997) has shown that there are some basic forms of deductive and empirical reasoning that we simply cannot do without. Like Descartes's *cogito* they keep bouncing back from every attempt to doubt them because every such attempt inevitably employs those same ways of reasoning. Thus, there are indeed some epistemic norms that are ahistorical and universal and underlie all rational inquiry (though I agree with Chalmers 1999, p. 171, that such epistemological commonplaces do not merit the grandiose name "Scientific Method"). Rescher argues that traditionally recognized forms of objective reasoning have a pragmatic rationale: "We can only achieve the essentially *pragmatic* justification of showing that, as best we can judge the matter, the counsels of reason afford the most promising systematic prospect of realizing our objectives" (Rescher 1997, p. 116; emphasis in original). Finally, some forms of reasoning can be justified naturalistically. Robin Dunbar has recently shown that many characteristically "scientific" methods, like the testing of hypotheses, are practiced by "prescientific" societies, small children, and even animals (Dunbar 1995). Some forms of reasoning just *come naturally,* and it is not clear that there is any practical alternative to relying upon them. There are therefore a number of ways to justify the basic forms of reason underlying scientific methods.

## 7. Beyond the Science Wars

1. See Peter Sacks's hilarious and horrifying *Generation X Goes to College* (1996). It details the ardors of attempting to educate the invincibly apathetic products of an educational system that preaches relativism and practices a pedagogy of spoon-feeding.

2. This has not always been the case. Darwin consciously attempted to base his work on the ideals of Herschel and Whewell, the leading philosophers of science in his day. In fact, the sharp boundaries that today sequester scholars into academic specialties were still blurred in Victorian times. For instance, T. H. Huxley rebutted St. George Mivart by an interpretation of the Latin and Greek Fathers of the Church. It is hard to imagine even the most humanistically inclined scientist of the present day accomplishing such a feat. More to the point, Darwin, Huxley, and others did not hesitate to engage in philosophical reflection on the implications of their science. For instance, the letters between Darwin and Asa Gray on the question of natural selection and divine design are far more philosophically sophisticated than many more recent discussions.

3. As noted in the introduction, by 1992 Bruno Latour claimed to have become a critic of social constructivism (Latour 1992). As David Hull notes in his review of Harry Collins and Trevor Pinch's *The Golem* (1993), the book is notable for the moderation of its claims (Hull 1995). Perhaps by the early 1990s the initial debunking fury of the constructivists had begun to wane.

4. Experience with readers of the manuscript has shown me that however

clearly I have tried to express this disclaimer, some will ignore it and regard me as an illicit intruder into the sociologists' domain. So I shall just say here what I think needs to be said and take my lumps.

    5. I am talking about the original (1954) cheap and lovable Japanese Godzilla, not the hideous Hollywood remake.

    6. Consider this passage from *An Introduction to Logic and Scientific Method* by Nagel and Morris Cohen (Cohen and Nagel 1934):

> Every experiment, therefore, tests not an isolated hypothesis, but the *whole body* of relevant knowledge logically involved. If the experiment is claimed to refute an isolated hypothesis, this is because the rest of the assumptions we have made are believed to be well founded. But this belief may be mistaken.... "Crucial experiments," we must conclude, are crucial against a hypothesis only if there is a relatively stable set of assumptions which we do not wish to abandon. But no guarantees can be given ... that some portion of such assumptions will never be surrendered. (pp. 220–21; emphasis in original)

These arguments, published in a textbook seventeen years before Quine's *Two Dogmas of Empiricism* (and, of course, traceable back to Pierre Duhem), show that the fragility and complexity of experimental results have long been recognized in the philosophy of science.

    7. Of course, the formation of the Chicxulub crater has not been directly observed by scientists. Its existence and features are inferred from core samples and mineralogical analysis (Glen 1994, p. 14; Frankel 1999). Such analysis shows that the Chicxulub rock is a high-iridium, glassy mineral (presumably formed from the cooling of the melt caused by impact) dating precisely to the K/T boundary. Other indirect evidence indicates that the Chicxulub "crater" is a 200 km wide circular structure off the tip of the Yucatan Peninsula (Glen 1994, p. 45). Some earth scientists still *strongly* question the credentials of the Chicxulub structure as a "smoking gun" of K/T mass extinction. See, for example, Officer and Page (1997) and N. MacLeod and G. Keller (1996).

    8. Lyell has recently received bad press, and his methodological prescriptions criticized as having retarded the growth of geology. For instance, Stephen Jay Gould says: "Lyell's gradualism has acted as a set of blinders, channeling hypotheses in one direction among a wide range of plausible alternatives.... Again and again in the history of geology after Lyell, we note reasonable hypotheses of catastrophic change, rejected out of hand by a false logic that brands them unscientific in principle" (Gould 1987, p. 176).

Gould's judgment is ahistorical and excessively harsh. As James Secord argues, Lyell's methodological recommendations were carefully qualified and were turned into rigid "isms" by later commentators (Secord 1997, p. xix). Lyell's uniformitarianism per se does not require absolute gradualism or preclude catastrophes. Humans had observed very substantial catastrophes, for example, the eruption of Tambora (1815) and the Lisbon earthquake (1755), so hypotheses postulating catastrophes of such a kind and intensity do not contravene Lyellian uniformity. Secord continues:

Lyell's public application of his principles was thus almost entirely regulative: that is, geologists should carry out their investigations *as though* visible causes are the same kinds as those that have acted in the past, and of the same degree of intensity. Uniformity of law, kind and degree had to be assumed, Lyell argued to make geology scientific. His subtitle cautiously spoke of "an *attempt*" to explain former changes; and the third and all subsequent editions softened this still further. . . . The positive aim was to make geology into an inductive science grounded in the observation of causes. (pp. xix–xx)

At bottom, Lyell's uniformitarianism seems to embody the eminently rational principle expressed in the saying "When you hear hoofbeats in the distance, think 'Ah, ha! Horses!' not 'Ah, ha! Unicorns!'" That is, try your best to explain phenomena in terms of ordinary causes before resorting to extraordinary ones. Of course any methodological maxim, however carefully qualified, can be turned into a dogma or applied in a ham-fisted way. Maybe this is Gould's real complaint, but Lyell can hardly be blamed for the subsequent misuse of his methodological recommendations.

9. In fact, impactors and gradualists *have* discovered considerable common ground. Even where they continue to disagree, the dispute is not over fundamental matters of principle like actualism versus catastrophism, but largely due to the differential application of practically learned judgmental skills to the various lines of evidence.

The excellent recent compendium *The Complete Dinosaur* contains a very interesting exchange between Dale A. Russell, a proponent of catastrophic extinction, and Peter Dodson, who still maintains a gradualist view (Russell and Dodson 1997, pp. 662–72). Their exchange indicates the broad areas of agreement that have emerged in the extinction debates of the last twenty years. Their continued disagreement over extinction hypotheses despite substantial agreement on the evidence is best explained by the inevitable uncertainties involved in practical judgments.

Russell and Dodson begin by listing thirteen points of agreement. Each side has made concessions to the other. For instance, they agree that the dinosaur record presently indicates a peak in global dinosaur diversity early in the Maastrichtian stage (the final stage of the Cretaceous Period), several million years before the end of the Cretaceous (p. 664). This is an important concession since one of the issues dividing the two camps has been whether dinosaurs had begun to decline well before the end of the Cretaceous or whether they had flourished until catastrophically destroyed. Further, both sides now also agree that it is reasonable to postulate a bolide strike coinciding with the final disappearance of the dinosaurs (p. 665). These mutual concessions show that the extinction debate, for all of its acrimony, did succeed in bringing closer together two initially polarized factions.

Russell and Dodson continue to disagree about how to weigh the various lines of evidence. Russell argues for abrupt changes in the end-Cretaceous biota:

> Apparently in both marine and terrestrial environments, the Cretaceous ended with an abrupt collapse in green plant productivity associated with the bolide trace-element signature. Marine and terrestrial animals belonging to food chains based on organic detritus tended to dominate the post-extinction assemblages. However,

> those dependent directly or indirectly on living plant tissues (e.g., dinosaurs on land and planktonic foraminifera and mosasaurs in the sea) are postulated to have died on time scales consistent with starvation. (p. 666)

He concludes:

> A relatively parsimonious interpretation of the foregoing points is that dinosaur dominated assemblages prospered in the Western Interior of North America until they were altered by regional topographic and climatic changes which began in the middle Maastrichtian time. Several million years after they had achieved a new balance regionally, these assemblages were decimated by a catastrophic environmental deterioration resulting from the impact of a comet. (ibid.)

Dodson sees more continuity than disruption across the K/T boundary and draws a different conclusion:

> The record of non-dinosaurian terrestrial and freshwater aquatic vertebrates, including fishes, amphibians, turtles, lizards, champosaurs, crocodiles, multituberculates, and placental mammals, shows substantial continuity across the Cretaceous-Tertiary boundary. . . . Plant communities also show continuity, although significant disruptions have been noted. These observations suggest that terrestrial communities did not suffer a devastating catastrophe, but responded to changing environmental conditions. (pp. 666–67)

Together, Russell and Dodson conclude:

> We have been pleasantly surprised to discover a broad area of common agreement. We concur that the latest Cretaceous dinosaurian record is far too incomplete to support either the catastrophic or the gradualistic model. . . . We differ in our assessment of which data are of greater significance. As we have described them, the two extinction models surely exist only in our imaginations. The truth lies in nature, which through the scientific method continually reveals ever-fascinating constellations of data which render the pursuit of scientific knowledge so enjoyable. (p. 669)

In short, the data are complex and conflicting and insufficient for decisive tests, so each scientist is thrown back on his own judgmental skills. For Russell, the indicators of catastrophe seem paramount; for Dodson, the continuities seem more striking than the disruptions. It is refreshing to see Russell and Dodson express themselves so modestly after the very strong rhetoric that has issued from both sides in this debate.

# REFERENCES

Ackrill, J. L., ed. 1987. *A New Aristotle Reader.* Princeton: Princeton University Press.
Alexander, R. M. 1976. "Estimates of Speeds of Dinosaurs." *Nature* 261: 129–30.
———. 1989. *Dynamics of Dinosaurs and Other Extinct Giants.* New York: Columbia University Press.
———. 1991. "How Dinosaurs Ran." *Scientific American* 264: 130–36.
Alvarez, L. W. 1987. *Alvarez: Adventures of a Physicist.* New York: Basic Books.
Alvarez, L. W., W. Alvarez, F. Asaro, and H. V. Michel. 1980. "Extraterrestrial Cause for the Cretaceous-Tertiary Mass Extinction." *Science* 208: 1095–1108.
Anonymous. 1912. " 'Dip' Plant on Full Time." *Pittsburgh Gazette Times,* February 20.
Archibald, J. D. 1996. *Dinosaur Extinction and the End of an Era.* New York: Columbia University Press.
———. 1997. "Extinction, Cretaceous." In P. J. Currie and K. Padian, eds., *Encyclopedia of Dinosaurs.* San Diego: Academic Press.
Ashworth, W. B., Jr. 1997. "Copernican Revolution." In J. Lankford, ed., *History of Astronomy: An Encyclopedia.* New York: Garland.
Asimov, A. 1966. *The History of Physics.* New York: Walker.
Avinoff, A. 1934. "Monthly Report of the Director of the Carnegie Museum for May, 1934." Archives of the Library of the Carnegie Museum of Natural History, Pittsburgh, Penn.
Bakker, R. T. 1968. "The Superiority of Dinosaurs." *Discovery* 3: 11–22.
———. 1971. "Dinosaur Physiology and the Origin of Mammals." *Evolution* 25: 636–58.
———. 1972. "Anatomical and Ecological Evidence of Endothermy in Dinosaurs." *Nature* 238: 81–85.
———. 1974. "Dinosaur Bioenergetics—A Reply to Bennett and Dalzell, and Feduccia." *Evolution* 28: 497–504.
———. 1975a. "Dinosaur Renaissance." *Scientific American* 232: 58–78.
———. 1975b. "Experimental and Fossil Evidence for the Evolution of Tetrapod Bioenergetics." In D. M. Gates and R. B. Schmerl, eds., *Perspectives of Biophysical Ecology,* pp. 365–99. New York: Springer.
———. 1980. "Dinosaur Heresy—Dinosaur Renaissance." In R. D. K. Thomas and E. C. Olson, eds., *A Cold Look at the Warm-Blooded Dinosaurs,* pp. 351–462. Boulder: Westview Press.
———. 1986. *The Dinosaur Heresies.* New York: William Morrow.
———. 1987. "The Return of the Dancing Dinosaur." In S. J. Czerkas and E. C. Olson, eds., *Dinosaurs Past and Present,* vol. 1. Los Angeles: Natural History Museum of Los Angeles County.
———. 1994. "The Bite of the Bronto." *Earth* 3, no. 6: 26–35.

# References

Bakker, R. T., and P. M. Galton. 1974. "Dinosaur Monophyly and a New Class of Vertebrates." *Nature* 248: 168–72.

Barnes, B. 1985. "Thomas Kuhn." In Q. Skinner, ed., *The Return of Grand Theory in the Human Sciences*. Cambridge: Cambridge University Press.

Barrick, R. E., M. K. Stoskopf, and W. J. Showers. 1997. "Oxygen Isotopes in Dinosaur Bone." In J. O. Farlow and M. K. Brett-Surman, eds., *The Complete Dinosaur*, pp. 474–90. Bloomington: Indiana University Press.

Baudrillard, J. 1988. "Simulacra and Simulations." In M. Poster, ed., *Jean Baudrillard: Selected Writings*. Stanford, Calif.: Stanford University Press.

Béland, P., and D. A. Russell. 1980. "Dinosaur Metabolism and Predator/Prey Ratios in the Fossil Record." In R. D. K. Thomas and E. C. Olson, eds., *A Cold Look at the Warm-Blooded Dinosaurs*, pp. 85–102. Boulder: Westview Press.

Bennett, A. F., and B. Dalzell. 1973. "Dinosaur Physiology: A Critique." *Evolution* 27: 170–74.

Benton, M. J. 1997. "Origin and Early Evolution of the Dinosaurs." In J. O. Farlow and M. K. Brett-Surman, eds., *The Complete Dinosaur*, pp. 204–15. Bloomington: Indiana University Press.

Berman, D. S., and J. S. McIntosh. 1978. "Skull and Relationships of the Upper Jurassic Sauropod Apatosaurus (Reptilia, Saurischia)." *Bulletin of the Carnegie Museum of Natural History* 8: 1–35.

Bernstein, R. J. 1983. *Beyond Objectivism and Relativism: Science, Hermeneutics, and Praxis*. Philadelphia: University of Pennsylvania Press.

Bloor, D. 1981. "Rational Reconstruction." In *Dictionary of the History of Science*, ed. W. F. Bynum, E. J. Browne, and R. Porter. Princeton: Princeton University Press.

———. 1991. *Knowledge and Social Imagery*. 2nd ed. Chicago: University of Chicago Press.

Bowler, P. J. 1992. *The Norton History of the Environmental Sciences*. New York: W. W. Norton.

Bretz, J. H. 1923a. "Glacial Drainage on the Columbia Plateau." *Bulletin of the Geological Society of America* 34: 573–608.

———. 1923b. "The Channeled Scabland of the Columbia Plateau." *Journal of Geology* 31: 617–49.

Brown, J. R. 1989. *The Rational and the Social*. London: Routledge.

Callebaut, W., ed. 1993. *Taking the Naturalistic Turn or How Real Philosophy of Science Is Done*. Chicago: University of Chicago Press.

Callon, M., and B. Latour. 1992. "Don't Throw the Baby Out with the Bath School! A Reply to Collins and Yearly." In A. Pickering, ed., *Science as Practice and Culture*. Chicago: University of Chicago Press.

Carnap, R. 1966. *An Introduction to the Philosophy of Science*. Edited by Martin Gardner. New York: Dover.

Cartmill, M. 1999. Review of *Mystery of Mysteries: Is Evolution a Social Construct? Reports of the National Center for Science Education* 19, no. 5: 49–50.

Chalmers, A. 1990. *Science and Its Fabrication*. Minneapolis: University of Minnesota Press.

———. 1999. *What Is This Thing Called Science?* 3rd ed. Indianapolis: Hackett.

Chapman, R. E. 1997. "Technology and the Study of Dinosaurs." In J. O. Farlow

# References

and M. K. Brett-Surman, eds., *The Complete Dinosaur*, pp. 112–35. Bloomington: Indiana University Press.

Chinsamy, A. 1993. "Bone Histology and Growth Trajectory of the Prosauropod Dinosaur *Massopondylus carinatus*." *Modern Geology* 18: 319–29.

Cohen, M. R., and E. Nagel. 1934. *An Introduction to Logic and Scientific Method*. New York: Harcourt, Brace.

Colbert, E. H. 1965. *The Age of Reptiles*. Mineola, N.Y.: Dover.

Collins, H. M. 1985. *Changing Order*. London: Sage.

——. 1994. "A Strong Confirmation of the Experimenter's Regress." *Studies in the History and Philosophy of Science* 25, no. 3: 493–503.

Collins, H. M., and T. Pinch. 1993. *The Golem: What Everyone Should Know about Science*. Cambridge: Cambridge University Press.

Craig, W. L. 1994. *Reasonable Faith: Truth and Christian Apologetics*. Wheaton, Ill.: Crossway Books.

Crompton, A. W., and S. M. Gatesy. 1989. Review of Gregory Paul, *Predatory Dinosaurs of the World*. *Scientific American* 260: 110–13.

Currie, P. J., and K. Padian, eds. 1997. *Encyclopedia of Dinosaurs*. San Diego: Academic Press.

Darwin, C. 1979. *The Origin of Species*. New York: Anvil Books.

Davis, M., P. Hut, and R. A. Muller. 1984. "Extinction of Species by Periodic Comet Showers." *Nature* 308: 715–17.

Davis, S. T. 1999. "Is Belief in the Resurrection Rational? A Response to Michael Martin." *Philo* 2, no. 1: 51–61.

Dennett, D. C. 1995. *Darwin's Dangerous Idea: Evolution and the Meanings of Life*. New York: Simon & Schuster.

De Ricqlés, A. J. 1980. "Tissue Structures of Dinosaur Bone." In R. D. K. Thomas and E. C. Olson, eds., *A Cold Look at the Warm-Blooded Dinosaurs*, pp. 103–39. Boulder: Westview Press.

De Salle, R., and D. Lindley. 1997. *The Science of Jurassic Park and the Lost World, or How to Build a Dinosaur*. New York: Basic Books.

Desmond, A. J. 1975. *The Hot-Blooded Dinosaurs: A Revolution in Paleontology*. New York: Dial Press.

——. 1982. *Ancestors and Archetypes: Paleontology in Victorian London 1850–1875*. Chicago: University of Chicago Press.

Dingus, L., and T. Rowe. 1997. *The Mistaken Extinction: Dinosaur Evolution and the Origin of Birds*. New York: W. H. Freeman.

Dodson, P. 1996. *The Horned Dinosaurs*. Princeton: Princeton University Press.

Douglass, E. Correspondence. Housed in the archives of the Department of Vertebrate Paleontology, the Carnegie Museum of Natural History, Pittsburgh, Penn. Transcribed and edited by Elizabeth Hill.

D'Souza, D. 1991. *Illiberal Education: The Politics of Race and Sex on Campus*. New York: Free Press.

Dunbar, R. 1995. *The Trouble with Science*. Cambridge, Mass.: Harvard University Press.

Eldredge, N. 2000. *The Triumph of Evolution and the Failure of Creationism*. New York: W. H. Freeman.

Farlow, J. O. 1980. "Predator/Prey Biomass Ratios, Community Food Webs and

# References

Dinosaur Physiology." In R. D. K. Thomas and E. C. Olson, eds., *A Cold Look at the Warm-Blooded Dinosaurs*, pp. 55–83. Boulder: Westview Press.

———. 1990. "Dinosaur Energetics and Thermal Biology." In D. B. Weishampel et al., eds., *The Dinosauria*, pp. 43–55. Berkeley: University of California Press.

Farlow, J. O., and M. K. Brett-Surman, eds. 1997. *The Complete Dinosaur.* Bloomington: Indiana University Press.

Farlow, J. O., and R. E. Chapman. 1997. "The Scientific Study of Dinosaur Footprints." In J. O. Farlow and M. K. Brett-Surman, eds., *The Complete Dinosaur*, pp. 505–18. Bloomington: Indiana University Press.

Farlow, J. O., and M. G. Lockley. 1991. "Ronald T. Bird, Dinosaur Tracker: An Appreciation." In D. D. Gillette and M. G. Lockley, eds., *Dinosaur Tracks and Traces*, pp. 43–55. Cambridge: Cambridge University Press.

Fastovsky, D. E., and D. B. Weishampel. 1996. *The Evolution and Extinction of the Dinosaurs.* Cambridge: Cambridge University Press.

Feduccia, A. 1973. "Dinosaurs as Reptiles." *Evolution* 27: 166–69.

Feyerabend, P. 1975. *Against Method.* London: NLB.

———. 1978. *Science in a Free Society.* London: NLB.

Fiorillo, A. R. 1997. "Taphonomy." In P. J. Currie and K. Padian, eds., *The Encyclopedia of Dinosaurs*, pp. 713–16. San Diego: Academic Press.

Forster, C. A., and P. C. Sereno. 1997. "Marginocephalians." In J. O. Farlow and M. K. Brett-Surman, eds., *The Complete Dinosaur*, pp. 317–29. Bloomington: Indiana University Press.

Frankel, C. 1999. *The End of the Dinosaurs: Chicxulub Crater and Mass Extinctions.* Cambridge: Cambridge University Press.

Franklin, A. 1994. "How to Avoid the Experimenter's Regress." *Studies in the History and Philosophy of Science* 25, no. 3: 463–91.

Gaffey, M. J. 1992. "Asteroids, Earth-Crossing." In S. P. Maran, ed., *The Astronomy and Astrophysics Encyclopedia.* New York: Van Nostrand Reinhold.

Gillette, D. D., and M. G. Lockley, eds. 1989. *Dinosaur Tracks and Traces.* Cambridge: Cambridge University Press.

Gilmore, C. W. 1936. "Osteology of *Apatosaurus*, with Special Reference to Specimens in the Carnegie Museum." *Memoirs of the Carnegie Museum* 11: 176–300.

Gish, D. T. 1992. *Dinosaurs by Design.* El Cajon, Calif.: Creation Life Publishers.

Glen, W. 1994. "Introduction," "What the Impact/Volcanism/Mass-Extinction Debates Are About," and "How Science Works in the Mass Extinction Debates," in W. Glen, ed., *The Mass Extinction Debates: How Science Works in a Crisis.* Stanford: Stanford University Press.

———. 1996. "Observations on the Mass Extinction Debate." *GSA Special Paper* 307, pp. 39–54.

Golinski, J. 1998. *Making Natural Knowledge: Constructivism and the History of Science.* Cambridge: Cambridge University Press.

Gould, S. J. 1987. *Time's Arrow, Time's Cycle: Myth and Metaphor in the Discovery of Geological Time.* Cambridge, Mass.: Harvard University Press.

———. 1991. *Bully for Brontosaurus.* New York: W. W. Norton.

Greenberg, N. 1980. "Physiological and Behavioral Thermoregulation in Living

# References

Pennock, R. T. 1999. *Tower of Babel: The Evidence against the New Creationism.* Cambridge, Mass.: MIT Press.

Pera, M. 1994. *The Discourses of Science.* Chicago: University of Chicago Press.

Porter, R. 1981. "Actualism" and "Uniformitarianism." In W. F. Bynum, E. J. Browne, and R. Porter, *Dictionary of the History of Science.* Princeton: Princeton University Press.

Psihoyos, L. 1997. "History of Dinosaur Discoveries: Research Today." In P. J. Currie and K. Padian, eds., *Encyclopedia of Dinosaurs,* pp. 352–55. San Diego: Academic Press.

Rainger, R. 1991. *An Agenda for Antiquity: Henry Fairfield Osborn and Vertebrate Paleontology at the American Museum of Natural History 1890–1935.* Tuscaloosa: University of Alabama Press.

Raup, D. M. 1981. "Probabilistic Models in Evolutionary Paleobiology." In B. J. Skinner, ed., *Paleontology and Paleoenvironments,* pp. 51–58. Los Altos, Calif.: William Kaufmann. Reprinted from *American Scientist* 65 (1977).

———. 1986. *The Nemesis Affair.* New York: W. W. Norton.

———. 1989. "The Case for Extraterrestrial Causes of Extinction." In W. G. Chaloner and A. Hallam, eds., *Evolution and Extinction.* Cambridge: Cambridge University Press.

———. 1991. *Extinction: Bad Genes or Bad Luck?* New York: W. W. Norton.

Raup, D., and J. J. Sepkoski. 1986. "Periodicity in Marine Extinction Events." In D. K. Elliott, ed., *Dynamics of Extinction.* New York: John Wiley & Sons.

Raup, D. M., and S. M. Stanley. 1978. *Principles of Paleontology.* 2nd ed. New York: W. H. Freeman.

Reid, R. E. H. 1987. "Bone and Dinosaurian 'Endothermy.'" *Modern Geology* 11: 133–54.

———. 1990. "Zonal 'Growth Rings' in Dinosaurs." *Modern Geology* 15: 19–48.

———. 1997. "Dinosaurian Physiology: The Case for 'Intermediate' Dinosaurs." In J. O. Farlow and M. K. Brett-Surman, eds., *The Complete Dinosaur,* pp. 448–73. Bloomington: Indiana University Press.

Rescher, N. 1997. *Objectivity: The Obligations of Impersonal Reason.* Notre Dame, Ind.: University of Notre Dame Press.

Riggs, E. S. 1903. "Structure and Relationships of Opisthocoelian Dinosaurs. Part I: *Apatosaurus* Marsh." *Publications of the Field Columbian Museum, Geology* 2: 165–96.

Romer, A. S. 1933. *Vertebrate Paleontology.* Chicago: University of Chicago Press.

———. 1945. *Vertebrate Paleontology.* 2nd ed. Chicago: University of Chicago Press.

———. 1956. *Osteology of the Reptiles.* Chicago: University of Chicago Press.

———. 1966. *Vertebrate Paleontology.* 3rd ed. Chicago: University of Chicago Press.

Roszak, T. 1968. *The Making of a Counter Culture.* Garden City, N.J.: Doubleday.

Rothbart, D. 1990. "Demarcating Genuine Science from Pseudoscience." In P. Grim, ed., *Philosophy of Science and the Occult,* 2nd ed., pp. 111–22. Albany: State University of New York Press.

# References

Ruben, J., et al. 1997. "New Insights into the Metabolic Physiology of Dinosaurs." In J. O. Farlow and M. K. Brett-Surman, eds., *The Complete Dinosaur*, pp. 505–18. Bloomington: Indiana University Press.

Rudwick, M. J. S. 1969. "Introduction." In C. Lyell, *Principles of Geology*, vol. 1, ix–xxv. New York: Johnson Reprint.

———. 1976. *The Meaning of Fossils*. 2nd ed. Chicago: University of Chicago Press.

———. 1990. "Introduction." In C. Lyell, *Principles of Geology*, vol. 1, pp. vii–lviii. Chicago: University of Chicago Press.

Ruse, M. 1976. "Charles Lyell and the Philosophers of Science." *British Journal for Philosophy of Science* 9, part 2, no. 32: 121–31.

———. 1979. *Sociobiology: Sense or Nonsense?* Dordrecht: Reidel.

———. 1995. Reply to letters in *The Sciences* 35, no. 2: 9.

———. 1999. *Mystery of Mysteries: Is Evolution a Social Construction?* Cambridge, Mass.: Harvard University Press.

Russell, D. A., and P. Dodson. 1997. "The Extinction of the Dinosaurs: A Dialogue between a Catastrophist and a Gradualist." In J. O. Farlow and M. K. Brett-Surman, eds., *The Complete Dinosaur*. Bloomington: Indiana University Press.

Sacks, P. 1996. *Generation X Goes to College*. Chicago: Open Court.

Sagan, C. 1977. *The Dragons of Eden: Speculations on the Evolution of Human Intelligence*. New York: Ballantine Books.

———. 1980. *Cosmos*. New York: Ballantine Books.

Scheffler, I. 1982. *Science and Subjectivity*. 2nd ed. Indianapolis: Hackett Publishing.

Schwartz, J. H. 1993. *What the Bones Tell Us*. New York: Henry Holt.

Searle, J. 1995. *The Construction of Social Reality*. New York: Free Press.

Secord, J. A. 1997. "Introduction." In C. Lyell, *Principles of Geology*, ed. J. A. Secord. London: Penguin Books.

Sepkoski, J. J. 1982. "A Compendium of Fossil Marine Families." *Milwaukee Public Museum Contributions to Biology and Geology* 51: 1–125.

Sepkoski, J. J., and D. Raup. 1986. "Periodicity in Marine Extinction Events." In D. K. Elliott, ed., *Dynamics of Extinction*. New York: John Wiley & Sons.

Sereno, P. 1991. Memoir 2. *Journal of Vertebrate Paleontology* 11 (supplement to no. 4).

———. 1999. "The Evolution of Dinosaurs." *Science* 284, no. 5423: 2137–47.

Shapin, S. 1982. "History of Science and Its Sociological Reconstructions." *History of Science* 20: 157–211.

———. 1992. Review of James Robert Brown, *The Rational and the Social*. *Philosophy of Science* 59: 712–13.

Shapin, S., and S. Schaffer. 1985. *Leviathan and the Air-Pump*. Princeton: Princeton University Press.

Shipman, P. 1981. *Life History of a Fossil*. Cambridge, Mass.: Harvard University Press.

———. 1998. *Taking Wing*. New York: Simon & Schuster.

Siegel, H. 1987. *Relativism Refuted*. Dordrecht: Reidel.

# References

Sismondo, S. 1996. *Science without Myth*. Albany: State University of New York Press.
Sokal, A. 1996. "Transgressing the Boundaries." *Social Text*, no. 46/47 (spring/summer): 217–52. Reprinted in Sokal and Bricmont 1998.
Sokal, A., and J. Bricmont. 1998. *Fashionable Nonsense: Postmodern Intellectuals' Abuse of Science*. New York: Picador.
Spalding, D. A. E. 1993. *Dinosaur Hunters*. Rocklin, Calif.: Prima Publishing.
Spotila, J. R. 1980. "Constraints of Body Size and Environment on the Temperature Regulation of Dinosaurs." In R. D. K. Thomas and E. C. Olson, eds., *A Cold Look at the Warm-Blooded Dinosaurs*, pp. 233–52. Boulder: Westview Press.
Stanley, S. M. 1986. *Extinction*. New York: Scientific American Books.
Taft, J. 1976. *Mayday at Yale: A Case Study in Student Radicalism*. Boulder: Westview Press.
Thomas, R. D. K., and E. C. Olson, eds. 1980. *A Cold Look at the Warm-Blooded Dinosaurs*. Boulder: Westview Press.
Thulborn, R. A. 1989. "The Gaits of Dinosaurs." In D. D. Gillette and M. G. Lockley, eds., *Dinosaur Tracks and Traces*, pp. 39–50. Cambridge: Cambridge University Press.
———. 1990. *Dinosaur Tracks*. London: Chapman and Hall.
Tornier, G. 1909. "Wie war der Diplodocus carnegii wirklich gebaut?" *Sitzungsbericht der Gesellschaft Naturforschender Freunde zu Berlin*, April 20, pp. 193–209.
Torrens, H. 1997. "Politics and Paleontology: Richard Owen and the Invention of Dinosaurs." In J. O. Farlow and M. K. Brett-Surman, eds., *The Complete Dinosaur*, pp. 175–90. Bloomington: Indiana University Press.
Toulmin, S. 1972. *Human Understanding: The Collective Use and Evolution of Human Concepts*. Princeton: Princeton University Press.
Van Fraassen, B.C. 1980. *The Scientific Image*. Oxford: Oxford University Press.
Varricchio, D. J. 1998. "Warm or Cold and Green All Over." *National Forum: The Phi Kappa Phi Journal* 78, no. 3: 16–20.
Wall, J. F. 1970. *Andrew Carnegie*. New York: Oxford University Press.
Ward, G. 1997. *Postmodernism*. Chicago: NTC/Contemporary Publishing.
Ward, P. D. 1998. *Time Machines: Scientific Explorations in Deep Time*. New York: Springer.
Weishampel, D. B., P. Dodson, and H. Osmolska. 1990. *The Dinosauria*. Berkeley: University of California Press.
Weissman, P. 1995. "Making Sense of Shoemaker-Levy 9." *Astronomy* 23, no. 5: 49–53.
Whitmire, D. P., and A. A. Jackson. 1984. "Periodic Comet Showers and Comet X." *Nature* 313: 36–38.
Wilde, C. B. 1981. "Whig History." In W. F. Bynum, E. J. Browne, and R. Porter, eds., *Dictionary of the History of Science*. Princeton: Princeton University Press.
Wilford, J. N. 1985. *The Riddle of the Dinosaur*. New York: Vintage Books.

# References

Woolgar, S. 1988. *Science: The Very Idea*. London: Tavistock Publications.
Worrall, J. 1988. "The Value of a Fixed Method." *British Journal for the Philosophy of Science* 39: 263–75.
Wright, W. 1998. *Born That Way: Genes, Behavior, Personality*. New York: Alfred A. Knopf.
Young, D. 1992. *The Discovery of Evolution*. Cambridge: Cambridge University Press.

# INDEX

AAAS Symposium on dinosaur endothermy, xvi, 26, 43, 98, 156
Actualism, 61, 62, 179n8
Alexander, R. McNeill, 135, 136, 142, 143
Algorithmic model of learning, 162. *See also* Canonical model of science
*Allosaurus,* xi
Alvarez, Luis, xii, xvii, 49–51, *50,* 60, 177n1, 179n9
Alvarez, Walter, 48, 49, *50,* 177n1, 179n9
American Museum of Natural History (AMNH), 1, 3, 4, 16, 116
Andrews, Roy Chapman, 122
Antirealism, 180–81
*Apatosaurus (Brontosaurus),* xi, xvii, *2,* 118;
  excavation of, 5–9;
  *louisae,* 10–12, *16;*
  name, 5–7;
  relation to *Diplodocus,* 14, 18–20;
  skull controversy, 2, 9–21, *16, 19,* 100, 101
*Archaeopteryx lithographica,* 126, 127, 129, 130, 184–85n6
Aristotle, 167, 168
*Atlantosaurus immanis,* 14
Avinoff, Andrey, 11, 12, 14

Baby Boomers, xi, 155
Bakker, Robert, *37;*
  on dinosaur bone histology, 35, 38;
  on dinosaur dominance, 23;
  as dinosaur "heretic," 22, 25, 98, 157;
  on dinosaur "orthodoxy," 23–25, 153;
  on dinosaur physiology, 26–38;
  on dinosaur posture and activity levels, 27–34, *30;*
  on dinosaur "renaissance," xi–xii;
  experiences that molded the views of, 153–56;
  on predator/prey ratios, 35, 36, 38–40, 42–43;
  use of rhetoric, 25, 152, 153
Barnes, Barry, 57, 58, 78, 177n3
Barrick, Reese, et al., 45
Baudrillard, Jean, 107–108
Behe, Michael, xiii

Béland, Pierre, 38, 39
Bennett, Albert F., 31, 32
Berman, David, 2, 17–21, 102
Bernstein, Richard, xxi, 158, 167–69
Blondlot, René-Prosper. *See* N-Rays
Bloor, David, 133, 147, 186n1
Bone histology. *See* Dinosaurs
Boyle, Robert, 138, 139, 146, 188n6
*Brachiosaurus,* 3, 118
Bretz, J. Harlan, 62
Bricmont, Jean, xiv
"Brontosaurus." *See Apatosaurus*
Brown, Barnum, 122
Brown, James Robert, 160, 161, 166
Butterfield, Herbert, 144

*Calvin and Hobbes,* 158
*Camarasaurus,* xvi, 2, 7, 11–21, *12, 13,* 100, 101
Canonical model of science, 162–64
Carnap, Rudolf, 162–63
Carnegie, Andrew, 1, 2, 3, 11, 21, 186n10
Carnegie Museum of Natural History, xvi, 2–5, 16
Carnegie Quarry, 4, 7, 11
Cartmill, Matt, 81–82
Catastrophist theories, 49, 51, 62
Chalmers, Alan, 160, 161, 166, 189n8, 190n8
Chicxulub crater, 172, 191n7
Cladistics, 137
CM11162 (fossil), 9, 10, 15–20, *16, 19*
*A Cold Look at the Warm-Blooded Dinosaurs* (Thomas and Olson), 26, 41, 43, 44, 98
Collins, Harry, xxi, 158–66, 169
Consensus, 80, 81
Constructivism. *See* Social Constructivism
Continental drift, 132, 187–88n4
"Conversion," 51, 52, 54, 58, 180n12. *See also* Raup, David
Cope, Edward Drinker, 5, 23, 115, *115*
Counterculture, 153–54
Craig, William Lane, 187
"Cynical" constructivism, 82, 83, 91

Dalzell, Bonnie, 31, 32
Darwin, Charles, 70, 96–98, 190n2
De Ricqlés, Armand J., 40–44
De Salle, Rob, 174

# Index

*Deinonychus*, 134
Dennett, Daniel, 125
Derrida, Jacques, 181
Desmond, Adrian, xx, 127–31
Dialogical model of science, 166–73
*Dimetrodon*, 36, 156
Dingus, Lowell, 134
Dinosaurs (*see also the names of particular genera*):
  activity levels, 28–30;
  bone histology, 35, 36, 38, 40, 41, 44;
  coloration, 117, 174;
  as cultural icons, xix, 106, 107, 112, 113;
  evolution, 27–29;
  extinction of (*see* K/T extinctions);
  gait, 28, 29;
  mass, 184n5;
  physiology, 24–47;
  reconstructions of, 23, 24, *24*, *30*, 114, *115*, *120*;
  speed, 136;
  taxonomy, 115, 116, 184–85n6;
  as totems, 109
*Diplodocus*:
  discovery of, 4;
  posture controversy, 118–21, *119*, *120*;
  skull of, 18, *19*, 100
Dodson, Peter, 29, 30, 192–93n9
Douglass, Earl, 4–9, *6*, 11, 122
Druyan, Ann, 124
Durham's Law, 39

Enculturational model of learning, 162
Endothermy, defined, 26, 27. *See also* Dinosaur: physiology
Experimenter's regress, 159–62, 165, 174
Extinction (*see also* K/T extinctions):
  Darwinian view of, 70–74;
  gradualist theories of, 51, 72;
  random models of, 72, 73;
  wanton extinction, 71, 72
*Extinction: Bad Genes or Bad Luck? See* Raup, David

Facts, social construction of, xviii, 81, 84–90
Farlow, James O., x, 24, 38–40, 43, 97
Feduccia, Alan, 29, 34
Feyerabend, Paul, 54, 56, 57, 140–42, 188–89n7
Franklin, Allan, 161, 162
Freud, Sigmund, 112, 113, 123
Froude number, 135–36
Fully Erect Gait, 28, 33

Gardner, Martin, 170
Gigantothermy, 45
Gilmore, C. W., 11, 14–17, 21, 25, 101
Gish, Duane, 90
Glen, William, 49, 52–57, 77, 78, 171, 178n4
Glen-Kuhn thesis, 56–59, 66, 67, 76, 180nn11,12
Godzilla, xi, 156, 191n5
Golinski, Jan, 82, 127, 131, 165
Gould, Stephen Jay, 191n8
Gradualism, 51, 66, 76
Grant, Robert, 128
Gravity waves, 158–62, 165
Greenberg, Neil, 42
Gross, Paul R., xiii–xiv, 91–93, 95

Hacking, Ian, 132
*Hadrosaurus*, 115
Hallam, A., 187n4
Haversian canals. *See* Dinosaurs: bone histology
Hanson, Russell Norwood, 84
Harvey, William, 89, 101
Hawkins, Benjamin Waterhouse, 114
Hay, Oliver P., 118, 119, *119*
Hellenius, Willem. *See* Ruben, John
Herschel, John, 61
*Higher Superstition. See* Gross, Paul R.
History of science, 127, 128, 131, 147–49
Hobbes, Thomas, 138
Hoffman, Antoni, 65, 172, 179n11
Holland, W. J., 2, 3, 7, 9–11, 13, 15, 17, 100, 118–21, *120*
Homeothermy, defined, 26. *See also* Dinosaur: physiology
Horner, John, xii
Hotton, Nicholas, III, 42
Hoyningen-Huene, Paul, 177n3
Hubbard, Ruth, 125
Huggett, Richard, 60
Hull, David, 164, 190n3
Huxley, T. H., xx, 23, 115, 126–31, *128*, 137, 144, 190n2

Icons, 106
Ideology, xii, 125, 128–31, 157
Impact theory of extinctions. *See* K/T extinctions
Incommensurability, xvii, 54–56, 168, 173
Iridium anomaly. *See* K/T extinctions

Ji, Qiang, et al., 44
Johnson, Phillip, xiii
*Jurassic Park*, 111

# Index

Kitcher, Phillip, 84, 103
Klee, Robert, 177–78n3, 178n5
Knight, Charles, 116, 117, 185–86n10
Kragh, Helge, 99, 100, 145, 146
K/T extinctions:
  impact theory of, 49, 50, 51–53, 172, 177nn1,2, 192–93n9;
  iridium anomaly, 53, 67, 179n9;
  "shocked minerals," 53, 172, 177n2;
  volcanic theory of, 51, 177n2
K/T horizon, 53, 179n9
Kuhn, Thomas, xvii, xxi, 51–58, 67, 79, 167, 168, 177n3, 178n5, 179–80n11. *See also* Glen-Kuhn thesis

*Laboratory Life*. *See* Latour, Bruno
LAGs (lines of arrested growth), 44
Lakatos, Imre, 141
*The Last Dinosaur Book*. *See* Mitchell, W. J. T.
Latour, Bruno:
  constructivist thesis, 90;
  "creeping realism," 93–95, 186n2;
  on dissident scientists, 91, 98;
  on the Enlightenment "faith," 93, 181n3;
  on "inversion," 89;
  *Laboratory Life*, xviii, 84–90;
  on networks, 91;
  on non-human "actants," 93, 95, 96;
  *The Pasteurization of France*, 91, 93, 101, 181–82n3;
  on "quasi objects," xix, 107;
  on "reality," xviii, 81, 82;
  on rhetoric, 88, 89, 99, 181n2;
  at the Salk Institute, 84–85;
  *Science in Action*, 91, 101;
  "warfare" model of science, 95, 98, 101–103
Laudan, Larry, 54, 132, 141, 144, 147
Laudan, Rachel, 54, 132
Leidy, Joseph, 115
Lennox, James, 71, 132
*Leviathan and the Air-Pump*. *See* Shapin, Steven
Levitt, Norman. *See* Gross, Paul
Lewis, Martin W., xiv
Lewontin, Richard, 125
Lindley, David, 174
Ludemann, Gerd, 187
Lull, R. S., 10
Lyell, Charles, 60, 61, 191–92n8

Marsh, O. C., 2, 5, 7, 10, 11, 13, 14, 18, 20, 23
Marx, Karl, 112

Matthew, W. D., 120
McGuire, Ted, ix, 86
McIntosh, John, 2, 17–21, 102
*Megalosaurus*, 115
Method, 140–43, 189–90n8. *See also* Paleontological practice and method
Mitchell, W. J. T.:
  on the coloration of dinosaurs, 117;
  on dinosaurs and capitalism, 122, 185–86n10;
  on dinosaur displays and male potency, 117, 121, 122;
  on dinosaur images, 114–17, 184n4;
  on dinosaurs as symbols, xix, 109, 110, 113, 183–84n2;
  fear of science, 124–25;
  on image and reality, 107, 108, 111, 112;
  on scientific vs. humanistic standards, 122, 183n1;
Morgan, J. P., 2, 3, 188n4
"Myth of rationalism," 57, 78

Nagel, Ernest, 163, 191n6
Nagel, Thomas, xv, 182–83n7, 190n8
Nasal turbinates, 45
*The Nemesis Affair*. *See* Raup, David
"Nemesis" hypothesis. *See* Raup, David
Newton-Smith, W. H., 54–56
Nonrationality Thesis (NT), 82
N-rays, 94–95

Objectivity, xviii, xix, 138, 182–83n7
Olby, Robert, ix, 179n8
Ornithischia, 116
Osborn, Henry Fairfield, 16, 100, 116, 117
Ostrom, John, xi, 91, 134
Owen, Richard, xx, 114, *115*, 126–31, *128*, 137, 144
Oxygen isotope ratios, 45

Padian, Kevin, 116, 184n4
Paleoichnology, 135, 142
Paleontological practice and method, 133–37, 139, 144, 188n5
Pera, Marcello, xviii, 96–98, 102, 103, 169
*Philosophy of Science*, 148
*Phronesis*. *See* Practical wisdom
Pinch, Trevor, 164
Popper, Karl, 142
Porter, Roy S., 61
Postmodernism, xviii, 107–109. *See also* Baudrillard, Jean
Practical wisdom, 167–70, 173, 174

# Index

Predator/prey ratios. *See* Bakker, Robert; Farlow, James O.
Progress. *See* Scientific progress
*Protoavis*, 100

Radical Meaning Variance, 56, 68, 69. *See also* Incommensurability
Rainger, Ronald, 16
Rational reconstruction, 141, 147
Rationalism, xiv, 57, 58, 80, 81, 82, 180n1, 189–90n8
Raup, David, xii, xvii, 64;
  "conversion" of, 58, 59, 66, 69, 78, 101, 179nn7,8, 180–81n12;
  on extinction, 71–76, 178n6, 179nn9,10;
  methods and standards of, 75, 76;
  "Nemesis" hypothesis, 63–65;
  on "philosophical" prejudices, 65–67, 171;
  response to Alvarez article, 60, 63, 67–68;
  work prior to "conversion," 59, 72–74
Reed, William H., 3, 4
Reid, R. E. H., 44, 46
Relativism, xiv, xv, 189–90n8
Relativism Thesis (RT), 82
Rescher, Nicholas, 143, 190n8
Rollins, Harold, ix, 178nn4,6
Romer, Alfred Sherwood, 15, 116
Roszak, Theodore, 152–54
Rowe, Timothy, 134
Ruben, John, and Willem Hellenius, 44–45
Rudwick, Martin, 53, 60, 61
Russell, Dale A., 38, 39, 192–93n9

Sagan, Carl, 124–25
Saurischia, 116
Sauropods. *See Apatosaurus; Diplodocus*
Schaffer, Simon, xx, 137–39, 143, 148, 149
*Science* (journal), xii, 49, 51, 60
Science and Technology Studies. *See* Sociology of scientific knowledge
Science wars, xiv, xv, xxii, 80, 83, 173
*Scientific Controversies* (Machamer, Pera, Baltas), 182n5
Scientific progress, 131–33, 144, 147
Secord, James A., 61, 191–92n8
Seeley, Harry Govier, 116
Semi-Erect Gait, 28
Sepkoski, J. John (Jack), Jr., 63, 65, 74, 75, 179n10, 179–80n11
Sereno, Paul, 34

Shapin, Steven, xx, 133, 137–39, 143, 148, 149
Shapin/Schaffer claim (S/Sc), 139–40
Shipman, Pat, 126
"Shocked" minerals. *See* K/T extinctions
Shoemaker/Levy 9 (comet), 172
Signor-Lipps Effect, 76
Sismondo, Sergio, 93, 95, 186n2
"Skeptical" constructivism, 82–83
Snow, C. P., 150
Social constructivism, xiv, xv, xxii, 80–83, 86–90, 103–105, 148. *See also* Facts, social construction of; Latour, Bruno
Sociology of scientific knowledge, xvi, xvii, xxi
Sokal, Alan, xiv, 85, 151, 152
Spotila, James R., 42
Sprawling Gait, 27
Stanley, Steven, 72
*Stegosaurus*, 28
Stewart, Douglas, 5, 7
*The Structure of Scientific Revolutions. See* Kuhn, Thomas

Taft, John, 152
Taphonomy, 134
Technocracy, 153
Thecodonts, 28, 36, 37
Therapsids, 30, 36
Thulborn, R. A., 188n5
Tornier, Gustav, 118–21, *119*, *120*
Toulmin, Stephen, 52, 78, 79, 169
*Triceratops*, xi, 28, 116
*Tyrannosaurus rex*, xi, 106, 108, 110, 116, 158, 174

Uniformitarianism, 60–62

*Velociraptor*, 156
Volcanic theory of K/T extinctions. *See* K/T extinctions

Ward, Glen, 107–108
"Warm blooded" dinosaur controversy. *See* Dinosaurs: physiology
Weber, Joseph, 158–61
Wegener, Alfred, 132
Whig History, xix, xx, 127, 128, 144–48
Williston, S. W., 14
Woolgar, Steve, xviii, 83. *See also* Latour, Bruno: *Laboratory Life*

Young, David, 104–105

**KEITH M. PARSONS** is Assistant Professor of Philosophy at the University of Houston, Clear Lake, and author of *God and the Burden of Proof.* He is former editor of *Philo,* Journal of the Society of Humanist Philosophers.